SpringerBriefs in Statistics

For further volumes:
http://www.springer.com/series/8921

Thomas W. MacFarland

Two-Way Analysis of Variance

Statistical Tests and Graphics Using R

 Springer

Thomas W. MacFarland
Office for Institutional Effectiveness
Nova Southeastern University
Fort Lauderdale, FL, USA
tommac@nova.edu

ISSN 2191-544X   e-ISSN 2191-5458
ISBN 978-1-4614-2133-7   e-ISBN 978-1-4614-2134-4
DOI 10.1007/978-1-4614-2134-4
Springer New York Dordrecht Heidelberg London

Library of Congress Control Number: 2011943305

Printed on acid-free paper

Springer is part of Springer Science+Business Media (www.springer.com)

Contents

Chapter 1
Learn R with Sample Lessons in Education and the Social Sciences, Health, and the Biological Sciences

Abstract R was developed in the early-to-mid 1990s and it has developed into a robust open source software environment, used with multiple operating systems for the organization, statistical analysis, and graphical presentation of data. As presented in this chapter, this set of lessons provides an introduction to the use of R and emphasizes good programming practices for data import or data entry, data organization, development of a detailed code book, visual data checks, descriptive analyses, and selected inferential analyses. Two-way analysis of variance (ANOVA) is the selected inferential test emphasized in this set of lessons, given how two-way ANOVA is well-suited to investigations that examine factorial designs that model complex real-world problems with practical applications. This chapter also introduces the multiple ways datasets are organized for use by R: tab-separated ASCII files, fixed width–fixed column ASCII files, and comma-separated values ASCII files. Throughout these discussions, graphical displays, and other actions that relate to quality assurance practices receive continual reinforcement.

1.1 Purpose of These Lessons

This set of lessons has been developed to take advantage of R and the many features available through this extensive and expanding open source environment. The purpose of this set of lessons is to provide directional guidance, with graphical reinforcement, for those who wish to use R to examine data, conduct statistical analyses, and present findings in graphical format.

Along with instruction on the use of R and R syntax associated with Two-Way analysis of variance (ANOVA), these lessons will also reinforce the use of descriptive statistics and graphical figures, to complement outcomes from two-way ANOVA. With attention to these lessons as well as other available resources, an interested student or beginning researcher should be able to use R to conduct and graphically display simple to complex statistical analyses. The R user community is growing, for good reason. R works, R is free, and R is fairly easy to learn.

T.W. MacFarland, *Two-Way Analysis of Variance: Statistical Tests and Graphics Using R*, SpringerBriefs in Statistics 1, DOI 10.1007/978-1-4614-2134-4_1,
© Thomas W. MacFarland 2012

1.2   Background on R

R is an open source software environment that is used for the organization, statistical analysis, and graphical presentation of data. Because R is offered as freeware, it is available at no direct monetary cost. In contrast, many proprietary software packages that provide similar or even fewer features may have an annual license cost of $1,000.00 USD or more.

R shares a heritage with both S and Scheme, previously developed programming languages that have useful applications for data organization and statistical analysis. The first iteration of the R environment was made available in the early-to-mid 1990s. Since then, the R user community has seen steady growth, with regular improvements to R and the concurrent development of supplemental library packages, expanding R far beyond its initial form. Updates to R are made frequently and are available through access to The Comprehensive R Archive Network (CRAN, http://cran.r-project.org/).

R is available for multiple operating systems, including Linux, Mac, UNIX, and Windows, with much parity and portability for all platforms. A long and growing set of supplementary packages for R is also available as freeware, to meet specific needs that are not included in the base package. As of September 2011, there were more than 3,000 R-specific packages available at no direct cost, merely by using resources posted at the CRAN Contributed Packages Website (http://cran.r-project.org/web/packages/).

1.3   Background on Two-Way ANOVA

A common statistical technique to determine if differences exist between multiple groups (however, the concept of group is defined) is one-way ANOVA and the associated *F* test. The *F* test and subsequent one-way ANOVA methodology involve the determination of differences for: (1) one group with multiple (typically, three or more) variations, as well as (2) one variable, compared to multiple (typically, three or more) groups.

Research designs are often far more complex, however, than merely determining if differences to a measured object variable exist from the perspective of one factorial (e.g., categorical) object variable. Consider a concept as complex as weight and, more specifically, weight gain of lab rats. Is weight gain an outcome of food intake (grams of food intake) only, or do categorical variables such as food quality (e.g., high protein diet, moderate protein diet, and low protein diet) and food palatability (e.g., high palatability, moderate palatability, and low palatability), etc. also impact weight gain?

To address this complexity, the two-way ANOVA statistical test is often used to determine differences (and possible interactions) when variables are presented from the perspective of two or more categories. When two-way ANOVA is used, it is possible to determine:

- Is there a difference because of variables acting independently of each other? In the above example, it would be useful to know if there is a difference in weight gain of lab rats because of (1) food quality or (2) food palatability.
- Is there a difference because of joint effects (e.g., interaction)? Again, in the above example, it would be useful to know if there is a difference in weight gain of lab rats because of interaction between (1) food quality and (2) food palatability.

Two-way ANOVA designs can become quite complex, not only to design but also to interpret. Yet, this highly useful methodology should not be avoided merely because it is not as simple as other statistical tests. On the contrary, two-way ANOVA should be used perhaps more than it is, due to the advantage of greater use of resources while modeling real-world scenarios.

Two-way ANOVA designs are often presented in a manner similar to other factorial analyses, such as the χ-square analysis. Like the χ-square analysis, two-way ANOVA uses a factorial organization with data placed in cells. The information within each cell provides the necessary data for later analysis. Thus, when using a two-way ANOVA, it is possible to examine three separate hypotheses: (1) Are the means for Variable A equal to the population? (2) Are the means for Variable B equal to the population? (3) Is there interaction between Variable A and Variable B?

To go back to the prior example of weight gain for lab rats, consider a scenario where (1) the measured object variable is weight gain (grams) of lab rats over an otherwise unspecified period of time, (2A) factorial object variable A represents the level of food quality (e.g., high protein diet, moderate protein diet, and low protein diet), and (2B) factorial object variable B represents the level of food palatability (e.g., high palatability, moderate palatability, and low palatability). In this example, a two-way ANOVA could be used to address the following questions:

- Is there a statistically significant difference in mean weight gain for the three protein levels? Intuitively, it would be reasonable to think that a high- protein diet would result in greater weight gain than would be experienced with a low-protein diet, but do the data support this assumption?
- Is there a statistically significant difference in mean weight gain for the three palatability levels? Intuitively, it would be reasonable to think that a highly palatable diet of easily digestible food would result in greater weight gain than would be experienced with a diet of low palatability, but do the data support this assumption?

Most importantly for the use of two-way ANOVA, is there any interaction between protein levels and palatability in terms of weight gain over time? Given the high cost of animal feed, it is best to recognize the possibility of interaction between food quality (high-protein feed is usually more expensive than low-protein feed) and food palatability (feed that is easily digestible is usually more expensive than feed that is of lower digestibility). With thousands of livestock typically kept at any given time in a feeder lot, even small differences in feed costs v weight gain can have a major impact on financial return and two-way ANOVA can help support investigations into possible pricing models.

Given this background, two-way ANOVA is often used to help us explain real-world scenarios, where interaction is often found, or at least possible. These more complex designs are different from simplistic designs that can only explain scenarios designed for simplistic modeling. The decision to use a two-way ANOVA is the decision to see if complex issues can be understood, and possibly acted upon.

1.4   Organization of These Lessons

There will be three samples in this R-based lesson on the use of two-way ANOVA. With each sample building on the other, the general approach in this lesson is to present increasingly detailed examples that show:

1.4.1   Data Import or Data Entry

Typically, import the data into the R session. Data are often put into comma-separated values (.csv) file format, but R can also import tab-separated data and data that are also in fixed width–fixed column format. Or, enter the data, if the dataset is small, directly into the R session.

1.4.2   Data Organization

Are numerical data actually codes for various factors, such as 1 represents Female and 2 represents Male? If so, it may be necessary to coerce the numerical data into factor format. Are fairly cryptic codes used to identify data, such as 1 represents Alachua County, FL, up to 67 represents Washington County, FL, with the use of numerical codes for an otherwise alphabetical listing of Florida's 67 counties? If so, it may be necessary to either create new object variables or at least better identify the codes so that the meaning of the data is presented in plain language, providing a meaningful description of the data and object variables. R provides many possibilities on how data can and should be organized, with functions such as `as.numeric()` and `is.numeric()` used to support desired organizational outcomes.

1.4.3   Display the Code Book

Are object variable names self-documenting? The use of 1 or 2 for Gender may be open to confusion without a code book. Does 1 represent Female and does 2 represent Male, or does 1 represent Male and does 2 represent Female? Codes such as F and M, for gender, may be a better attempt at self-documentation, but even with these codes there may be a misunderstanding without an explicit set of declarations in a code book. Will lowercase f be accepted as an alternate presentation of uppercase F and will lowercase m be accepted as an alternate presentation of uppercase M? However, the code book is constructed, the codes should always be clearly detailed. It is also reasonable to identify expected ranges for *high* and *low* values, when appropriate, to help identify later values that may be either illogical or out-of-range. Even when the data seem self-explanatory, remember that data are often shared with others and as such, others may not have sufficient background knowledge to know all subtleties about the data and background surrounding the data. Consider the term Washington County as a datum. Does the term Washington County refer to Washington County in Alabama, Arkansas, Colorado, Florida, Georgia, Idaho, Illinois, Indiana, Iowa, Kansas, Kentucky, Maine, Maryland, Minnesota, Mississippi, Missouri, Nebraska, New York, North Carolina, Ohio, Oklahoma, Oregon, Pennsylvania, Rhode Island, Tennessee, Texas, Utah, Vermont, Virginia, Wisconsin, or Washington Parish, Louisiana? Self-documentation is useful, but explicit documentation is more prudent.

1.4.4   Conduct a Visual Data Check

Graphics are an excellent way to check the data for values that are either illogical or out-of-range. All graphical images do not have to be of publishable quality. It is common, and encouraged, to use simple graphical images to visually review outcomes and to add another component to data quality assurance. Initially, a simple graphic in black and white with minimal detail may be all that is needed to visually check the data for illogical or out-of-range data. Then, if appropriate, a more detailed graphic of publishable quality can always be generated.

1.4.5   Descriptive Analysis of the Data

Descriptive statistics such as N, Mean, Standard Deviation (SD), Median, Range, etc., provide an excellent first view of the data. It would be the rare attempt at data analysis that did not include a full set of descriptive analyses, with the emphasis on statistics associated with measures of central tendency. Summary descriptive statistics provide an overall view of the data, with quantitative description of the data at the broadest level. Breakout descriptive statistics provide even greater detail than summary descriptive statistics. As a typical example, when working with livestock and other biological specimens (including humans), it is often important to consider differences between the two genders, Female and Male. Breakout descriptive statistics are used to provide this level of detail.

Because these many activities require a fair amount of planning and time-on-task, it is essential to have a dataset that is correctly organized and understood. The outcomes of later statistical analyses can only be accepted if there is prior acceptance and understanding of the dataset. Indeed, it is not at all uncommon for experienced researchers to devote more time to dataset quality assurance tasks than the time devoted to analytics. Quality outcomes are best gained through attention to quality inputs.

1.5   Details of the Three Sample Datasets

1.5.1   Tab-Separated ASCII File

The first dataset is a collection of Software Engineering Final Examination scores (0–100) by (A) teaching method (1 = Lecture, 2 = CBT (Computer-based training), 3 = Video, 4 = IDS (independent study) and by (B) status as a Community College graduate (1 = is a Community College graduate, 2 = is not a Community College graduate). The dataset consists of fewer than 100 subjects and there are no missing data. The first dataset was prepared as a tab-separated ASCII file.

1.5.2   Fixed Width–Fixed Column ASCII File

The second dataset is a collection of measured data (e.g., height, weight, blood pressure, etc.) and responses to a Wellness Inventory (e.g., smoking habits, drinking habits, etc.). The dataset consists of more than 100 subjects and typical to the realities of a field-based activity where possibly intrusive questions and measurements are involved, data for a few individual responses are missing. The second dataset was prepared as a fixed width–fixed column ASCII file.

1.5.3   Comma-Separated Values ASCII File

The third dataset is a collection of data taken from an agricultural integrated pest management (IPM) study. Data are all numeric and represent the three object variables of interest to this study: formulation of the agricultural chemical (three factors), time-of-day for chemical application (two factors), and the number of larvae for a specific insect per square meter at random locations one week after chemical application. The dataset has no missing data and there is an equal number of measurements for each cell. The third dataset was prepared in .csv format, as a comma-separated values ASCII file.

Presentation of how R is used against these three datasets goes from simple to detailed. The first dataset, based on a student learning outcomes assessment in Education, is examined in a fairly simple manner. Complete analyses and useful graphics are provided, but in an attempt to provide simple examples, complexity is kept to a minimum. The second dataset, based on a Health Science Wellness Inventory, introduces a moderate degree of complexity. There are some challenges to the dataset and graphical presentations increase in detail and complexity. The third dataset, using data from the Biological Sciences, is examined in complete detail. Details are examined closely and graphical presentation approaches publishable quality. By following along with this measured level of increased detail, self-confidence in the use of R is developed and new R functions are introduced with each lesson. Be sure to examine all three lessons.

Chapter 2
Two-Way Analysis of Variance (ANOVA) Sample 1: Comparison of Scores on a Final Examination by Teaching Method and by Status as a Community College Graduate

Abstract R is used in this chapter to support investigations on the assessment of student learning outcomes in higher education. A tab-separated file was used as the data source and from this file it was demonstrated how R is used to organize and label data, prepare simple graphical figures for quality assurance purposes, provide descriptive statistics overall and by breakout groups, and conduct a two-way analysis of variance (ANOVA). This chapter provides a simple introduction on the use of R, with an emphasis on easy-to-follow confidence building actions that model a real assessment-type setting faced daily by those who work in education and the social sciences.

This study was designed to examine if there are differences in final examination test scores for students in a software engineering course by teaching method (four breakout groups: (1) traditional lecture (e.g., lecture), (2) computer based training (e.g., CBT), (3) instructional video, placed on a tablet computing device (e.g., video), and (4) independent study (e.g., IDS)) and by Status as a Community College graduate (two breakout groups: (1) graduate of a Community College and (2) not a graduate of a Community College).

The motivation for this study was to react to university-level faculty concerns about the skills two-year community college graduates bring into advanced courses (overall and by different teaching methods), as compared to the skills of students who enroll in their first two years of undergraduate study at a four-year university. For background on methods, students were all enrolled in a university senior-level software engineering course. Students were assigned, through an alpha-sort by last name random selection process, to placement into one of the four teaching method groups. Community College graduation status was determined by transcript data maintained in the Registrar's Office. All students sat for the same final examination. This design will allow investigations by teaching method, by Community College graduation status, and by possible interactions between these two factor-type object variables (teaching method and Community College graduation status).

T.W. MacFarland, *Two-Way Analysis of Variance: Statistical Tests and Graphics Using R*, SpringerBriefs in Statistics 1, DOI 10.1007/978-1-4614-2134-4_2,
© Thomas W. MacFarland 2012

9

The principal investigator was confident that final examination scores represented interval data. As such, two-way analysis of variance (ANOVA) was judged to be the appropriate test for this factorial-type analysis of summative differences in final examination scores by teaching method (four breakout groups) and by Status as a Community College graduate (two breakout groups). The header, the first three lines of data, and the last three lines of data, presented in tab-separated format, are shown below, but of course the tab characters although present do not show:

```
ID Method ComCol Final
01 1 1 089
02 1 1 081
03 1 2 073
73 4 2 062
74 4 1 056
75 4 1 085
```

Sample 1 Ho (Null Hypothesis): There is no difference between teaching method, graduation status from a Community College, and interaction between teaching method and graduation status from a Community College regarding final examination test scores of students enrolled in a university senior-level software engineering course ($p <= 0.05$).

2.1   Data Import of a .txt Tab-Delimited Data File into R

For this lesson, notice how the dataset has been prepared in .txt file format (not .csv file format) and that the data are separated by tab spacings, not commas. Experienced researchers will work with data in many formats, so it is desirable to gain experience with this type of file format even though .csv (comma-separated values) data files are certainly quite common when using R and other data analysis tools. The data for this lesson, in .txt tab-delimited format, have been placed at the `F:\R_Lessons\Inferential_Statistics_Parametric` directory on a standalone personal computer.

All analyses begin from this starting point, working with a previously prepared .txt file. The emphasis will be on: (1) overall analysis of Final, (2) breakout analyses of Final; Final by teaching method and Final by Status as a Community College graduate, (3) graphical representation of overall and breakout findings, (4) two-way ANOVA of the data, and (5) summative interpretation of outcomes.

```
################################################################
# Housekeeping                          Use for all analyses
################################################################
setwd("F:/R_Lessons/Inferential_Statistics_Parametric")
                    # Set to a new working directory.
                    # Note the single forward slash and double
                    # quotes.
```

```
                        # This new directory should be the directory
                        # where the data file is located, otherwise
                        # the data file will not be found.
getwd()                 # Confirm the working directory.
search()                # Attached packages and objects.
##############################################################
```

2.1.1   Data Import or Data Entry

```
Final.table <- read.table (file =
  "SoftwareEngineeringFinal_PriorCC_tab-separated.txt",
  header = TRUE,
  sep = "\t")            # Import the tab-separated .txt
                         # file.
getwd()                  # Identify the working directory.
ls()                     # List objects.
attach(Final.table)      # Attach the data, for later use.
names(Final.table)       # Identify names.
head(Final.table)        # Show the head.
tail(Final.table)        # Show the tail.
Final.table              # Show the entire data frame.
```

By completing this action, an object called Final.table has been created. This object consists of the data included in the tab-separated file. Make sure that there are no prior R-based datasets called Final.table available. Note how it was only necessary to key the filename for the .txt file and not the full pathname since the R working directory is currently set to the directory and subdirectory where this .txt file is located (see Sect. 2.1 at the beginning of this lesson).

2.2   Organize the Data and Display the Code Book

Now that the data have been imported into R, it is usually necessary to check the data for format and then make any changes that may be needed, to organize the data. As a typical example, consider the common practice of numeric codes, as factors, for Gender. If Gender is coded as 1 and 2 instead of Female (1) and Male (2), it is necessary to do something so that 1 and 2 are seen as factor (e.g., group) values and not integers more suited for math operations. This concept applies to all other cases where numeric codes are used to identify factors (e.g., groups). In this example, that concept applies to the numeric codes used to identify the four teaching methods and the two options regarding status as a Community College graduate.

```
class(Final.table)
class(Final.table$ID)       # DataFrame$ObjectName notation.
class(Final.table$Method)   # DataFrame$ObjectName notation.
```

```
class(Final.table$ComCol)     # DataFrame$ObjectName notation.
class(Final.table$Final)      # DataFrame$ObjectName notation.
```

Now that the class() function has been applied against each object, consult the code book and coerce each object, as needed, into its correct class. Typically, integers that serve as numeric codes (e.g., 1 represents Female and 2 represents Male) are coerced into factor format.

```
# Code Book ##############################################
##########################################################
# Software Engineering Final Examination Results by
  Teaching Method and by Status as a Community
  College Graduate
#
# Variable Labels
#   ID          Student Identification Number
#   Method      Teaching Method
#   ComCol      Status as a Community College Graduate
#   Final       Final Examination Score
#
# Variable Values
#   ID          Nominal    LOW to HIGH
#   Method      Nominal    1 Lecture
#                          2 CBT (Computer-based Training)
#                          3 Video
#                          4 IDS (Independent Study)
#   ComCol      Nominal    1 Yes Is a Community College
#                                Graduate
#                          2 No  Is not a Community
#                                Community Graduate
#   Final       Interval   000 Lowest Possible Score
#                          100 Highest Possible Score
##########################################################
```

In an effort to promote self-documentation and readability, it is often desirable to label all object variables. The epicalc::label.var() function can serve this purpose. Of course, be sure to load the epicalc package, if it is not operational from prior analyses.

```
install.packages("epicalc")
library(epicalc)          # Load the epicalc package.
help(package=epicalc)     # Show the information page.
sessionInfo()             # Confirm all attached packages.
```

Comment: Use help(library) and help(require) to see the functional difference, if any, between these two functions.

```
epicalc::des(Final.table)
```

Use the `epicalc::des()` function to see the nature of the data frame. Then, provide a useful description of each object variable by using the `epicalc::label.var()` function.

```
epicalc::label.var(ID,      "Subject ID",
  dataFrame=Final.table)
epicalc::label.var(Method, "Teaching Method",
  dataFrame=Final.table)
epicalc::label.var(ComCol, "Community College Status",
  dataFrame=Final.table)
epicalc::label.var(Final,  "Final Examination Score",
  dataFrame=Final.table)


epicalc::des(Final.table)
  # Confirm the description of each object variable.
```

Coerce objects into correct format. Notice how variables are named: `DataFrame$ObjectName`. At first this action may seem somewhat cumbersome, but it is actually very useful to ensure that actions are performed against the correct object. Most text editors allow the use of copy/paste and find/replace, so it should be a simple operation to organize the syntax.

```
# Object class before coercion
class(Final.table)
class(Final.table$ID)        # DataFrame$ObjectName notation.
class(Final.table$Method)    # DataFrame$ObjectName notation.
class(Final.table$ComCol)    # DataFrame$ObjectName notation.
class(Final.table$Final)     # DataFrame$ObjectName notation.

# Coercion
Final.table$ID       <- as.factor(Final.table$ID)
Final.table$Method   <- as.factor(Final.table$Method)
Final.table$ComCol   <- as.factor(Final.table$ComCol)
Final.table$Final    <- as.numeric(Final.table$Final)

# Object class after coercion
class(Final.table)
class(Final.table$ID)        # DataFrame$ObjectName notation.
class(Final.table$Method)    # DataFrame$ObjectName notation.
class(Final.table$ComCol)    # DataFrame$ObjectName notation.
class(Final.table$Final)     # DataFrame$ObjectName notation.
```

As a sidebar comment, at the R prompt, key help(as.numeric) and then key help(as.integer) to see the differences between these two R functions and when it may be best to use each.

Use the `str()` function to confirm object format. Note the details for `str()` output, especially the output against the data frame Final.table.

```
str(Final.table)
str(Final.table$ID)
```

```
str(Final.table$Method)
str(Final.table$ComCol)
str(Final.table$Final)
```

The epicalc package has many useful functions. Saying this, use the `epicalc::des()` function to again describe the data frame currently in use.

```
epicalc::des(Final.table)
```

Then, merely to further confirm the nature of the dataset, use the `levels()` function against the factor object variables, to reinforce understanding of the data.

```
levels(Final.table$Method)
levels(Final.table$ComCol)
```

Use the `summary()` function against the object Final.table, which is a data frame, to gain an initial sense of descriptive statistics and frequency distributions.

```
summary(Final.table)
```

Although the dataset seems to be in correct format, it is somewhat difficult to work with numeric values for factor object variables: Method and ComCol. Use the code book to review the meaning for each factor code and then note how this problem is easy to accommodate. As is nearly always the case with R, there is no one-and-only-one way to apply labels and recode data.

```
# Apply the labels() function.
Final.table$Method.recode   <- factor(Final.table $Method,
   labels = c("Lecture", "CBT", "Video", "IDS"))

head(Final.table$Method)
head(Final.table$Method.recode) # View the first lines of data.

# Apply the labels() function.
Final.table$ComCol.recode   <- factor(Final.table $ComCol,
   labels = c("Yes - is a CC Graduate",
              "No  - not a CC Graduate"))

head(Final.table$ComCol)
head(Final.table$ComCol.recode) # View the first lines of data.
```

To recap, the object variable `Final.table$Method.recode` was created by applying the `factor()` function against the object variable `Final.table$Method`. The `labels()` function was used to embellish future output into simple English. This same approach was also used to create the object variable `Final.table$ComCol.recode` and then create labels for this new object variable.

Now, merely use the `attach()` function again to confirm that all data are attached to the data frame.

```
attach(Final.table)
head(Final.table)
tail(Final.table)
summary(Final.table)   # Quality assurance data check.
str(Final.table)       # List all objects, with finite detail.
```

As an additional data check, use the `table()` function to see how data have been summarized using the newly created names (factor object variables) as well as the original names for the numeric object variables.

```
table(Final.table$Method,         useNA = c("always"))
table(Final.table$Method.recode, useNA = c("always"))

table(Final.table$ComCol,         useNA = c("always"))
table(Final.table$ComCol.recode, useNA = c("always"))

table(Final.table$Method.recode,
      Final.table$ComCol.recode, useNA = c("always"))
```

Note how the argument `useNA = c("always")` is used with the table function, to force identification of missing values.

This type of redundancy and attention to detail at this stage of development may seem unnecessary, but it more than helps in reducing later errors caused by a simple oversight.

2.3   Conduct a Visual Data Check

The `summary()` function, `min()` function, and `max()` function are all certainly useful for data checking, but there are also many advantages to a visual data check process. In this case, simple plots and bar charts can be very helpful in an attempt to look for data that may be either illogical or out-of-range. These initial plots and barcharts will be, by design, simple, and should be considered throwaways as they are intended only for initial diagnostic purposes. They will then be followed by graphical images that provide more detail and may have future use in any possible presentation(s).

```
names(Final.table)    # Confirm all object variables.
```

2.3.1   Simple Plots

```
par(ask=TRUE)
plot(Final.table$ID,
  main="Final.table$ID Visual Data Check",
```

```
col = c(rainbow(75)))
# rainbow refers to the 75 datapoints for ID.

par(ask=TRUE)
plot(Final.table$Method,
  main="Final.table$Method Visual Data Check",
  col = c(heat.colors(4)))
# heat.colors refers to the 4 groups for Method.

par(ask=TRUE)
plot(Final.table$Method.recode,
  main="Final.table$Method.recode Visual Data Check",
  col = c(terrain.colors(4)))
# terrain.colors refers to the 4 groups for Method.recode.

par(ask=TRUE)
plot(Final.table$ComCol,
  main="Final.table$ComCol Visual Data Check",
  col = c(topo.colors(2)))
# topo.colors refers to the 2 groups for ComCol.

par(ask=TRUE)
plot(Final.table$ComCol.recode,
  main="Final.table$ComCol.recode Visual Data Check",
  col = c(cm.colors(2)))
# cm.colors refers to the 2 groups for ComCol.recode.

par(ask=TRUE)
plot(Final.table$Final,
  main="Final.table$Final Visual Data Check",
  pch=19,
  col = c("black"))
# pch=19 displays datapoints as black solid circles.
```

The purpose of these initial plots is to gain a general sense of the data and to equally look for outliers. In an attempt to look for outliers, the ylim argument has been avoided, so that all data are plotted. Extreme values may or may not be outliers, but they are certainly interesting and demand attention.

2.3.2 Histogram of the Summary Object Variable

This sample lesson has been designed to look into the nature of object variable Final and the factor object variables Method and ComCol, recoded into a more verbose format as object variables Method.recode and ComCol.recode. Given the nature of Final values, it may also be a good idea to supplement the plot() function with

other functions, to gain a different view of the continuous values of Final, overall
and by breakout groups.

```
par(ask=TRUE)
hist(Final.table$Final,
  main="Final.table$Final Visual Data Check(Histogram)",
  font=2, cex.lab=1.15, col="red")
```

For questions about this function and all other functions, simply key
help(function.name) to learn more about the R function and the many
arguments and options supported by the function. The use of arguments and
attention to axis scales can greatly improve presentation, as shown in the next
set of syntax.

```
par(ask=TRUE)
hist(Final.table$Final,
  main="Histogram of Final Exam Values",
  xlab="Final Exam Values (Limit = 0 to 100)",
  ylab="Frequency",
  xlim=c(0,120),    # Note the selection for xlim.
  ylim=c(0,25),     # Note the selection for ylim.
  cex.lab=1.15, cex.axis=1.15, freq=TRUE,
  border="blue", col="red")
# Again, note the xlim=c(0,120) argument. By design,
# the X axis is pushed out to 120 instead of 100, to
# accommodate a full presentation of output.
```

Compare the output of the histogram to the normal curve and decide if overall
distribution allows the use of planned inferential tests. Again, many tests are
sufficiently robust to allow their use even if normal distribution is desired, but not
fully met.

2.3.3   Horizontal and Vertical Boxplots of the Summary Object Variable

The boxplot is another tool typically used to gain a more complete understanding of
the values represented by an object variable. Largely depending on preference and
need, a boxplot can be displayed in either horizontal or vertical orientation while
still highlighting the lower quartile (Q1), the median (Q2), the upper quartile (Q3),
and any possible outliers.

```
par(ask=TRUE)
boxplot(Final.table$Final,
  horizontal=TRUE,
  main="Horizontal Boxplot of Final Exam Values",
```

```
  xlab="Final Exam Values (Limit = 0 to 100)",
  ylim=c(0,125),      # Note the selection for ylim.
  cex.lab=1.15, cex.axis=1.15,
  lty=1,              # Note the line type.
  lwd=3,              # Note the line width.
  border="blue", col="red")
box()

par(ask=TRUE)
boxplot(Final.table$Final,
  horizontal=FALSE,
  main="Vertical Boxplot of Final Exam Values",
  ylab="Final Exam Values (Limit = 0 to 100)",
  ylim=c(0,125),      # Note the selection for ylim.
  cex.lab=1.15, cex.axis=1.15,
  lty=1,              # Note the line type.
  lwd=3,              # Note the line width.
  border="blue", col="red")
box()
```

In the same way that descriptive statistics at the breakout level provide detail about selected object variables for specific groups (e.g., details on Female subjects v details on Male subjects), graphical displays at the breakout level are also useful. It is especially helpful to graphically display object variables at the breakout level to compare distribution, extreme values, etc. This is all done in an effort to determine, later, if there are (or are not) differences by breakout groups.

2.3.4   Sorted Dot Chart of the Summary Object Variable by Breakout Object Variables

```
par(ask=TRUE) #Method.recode
epicalc::summ(Final.table$Final,
  by=Final.table$Method.recode,
  graph=TRUE, # Use graph=TRUE, if desired.
  pch=20, ylab="auto",
  main="Sorted Dotplot of Final Exam Values
  by Teaching Method",
  cex.X.axis=1.25, # Note X axis label size.
  cex.Y.axis=1.25, # Note Y axis label size.
  font.lab=2, dot.col="auto")
```

```
par(ask=TRUE) # ComCol.recode
epicalc::summ(Final.table$Final,
  by=Final.table$ComCol.recode,
  graph=TRUE, # Use graph=TRUE, if desired.
  pch=20, ylab="auto",
  main="Sorted Dotplot of Final Exam Values
  by Community College Graduation Status",
  cex.X.axis=1.25, # Note X axis label size.
  cex.Y.axis=0.95, # Note Y axis label size.
  font.lab=2, dot.col="auto")
```

These sorted dot charts provide a clear view of final exam values by teaching method and then by Community College graduation status. The epicalc::summ() function and the accompanying graph=TRUE argument should always be among the first graphical tools used to examine distributions by breakout group.

2.3.5   Histogram of the Summary Object Variable by Breakout Object Variables

It is certainly possible to put all final exam values for Lecture into a separate object and to then prepare a histogram of only lecture-based final exam values. This action could then be repeated so that there are separate object variables for all teaching methods: lecture, CBT, video, and IDS. By using the lattice package, however, this excessively redundant action can be avoided, with separate histograms for each teaching method placed into the same graphical image. This action can then be repeated for the two Community College graduation status groups: Yes and No. This general approach and subsequent use of the lattice package can be used for other graphic techniques, as needed.

Breakout Histograms by Count

```
install.packages("lattice")
library(lattice)          # Load the lattice package.
help(package=lattice)     # Show the information page.
sessionInfo()             # Confirm all attached packages.

par(ask=TRUE) # Method.recode
lattice::histogram(~Final.table$Final |
                    Final.table$Method.recode,
  type="count",          # Note:  count
  par.settings=simpleTheme(lwd=2),
  par.strip.text=list(cex=1.15, font=2),
```

```
   scales=list(cex=1.15),
   main="Histograms (Count) of Final Exam Values
   by Teaching Method",
   xlab=list("Final Exam Values (Limit = 0 to 100)",
   cex=1.15, font=2),
   xlim=c(0,120),
   ylab=list("Count", cex=1.15, font=2),
   aspect=0.2,
   layout = c(1,4),  # Note:  1 Column by 4 Rows.
   col="red")

par(ask=TRUE) # ComCol.recode
lattice::histogram(~Final.table$Final |
                    Final.table$ComCol.recode,
   type="count",        # Note:  count
   par.settings=simpleTheme(lwd=2),
   par.strip.text=list(cex=1.15, font=2),
   scales=list(cex=1.15),
   main="Histograms (Count) of Final Exam Values by
   Community College Graduation Status",
   xlab=list("Final Exam Values (Limit = 0 to 100)",
   cex=1.15, font=2),
   xlim=c(0,120),
   ylab=list("Count", cex=1.15, font=2),
   aspect=0.2,
   layout = c(1,2),  # Note:  1 Column by 2 Rows.
   col="red")
```

Breakout Histograms by Percent

```
par(ask=TRUE) # Method.recode
lattice::histogram(~Final.table$Final |
                    Final.table$Method.recode,
   type="percent",      # Note:  percent
   par.settings=simpleTheme(lwd=2),
   par.strip.text=list(cex=1.15, font=2),
   scales=list(cex=1.15),
   main="Histograms (Percent) of Final Exam Values by
   Teaching Method",
   xlab=list("Final Exam Values (Limit = 0 to 100)",
   cex=1.15, font=2),
   xlim=c(0,120),
   ylab=list("Percent", cex=1.15, font=2),
   aspect=0.2,
```

```
layout = c(4,1),  # Note:  4 Columns by 1 Row.
col="red")

par(ask=TRUE) # ComCol.recode
lattice::histogram(~Final.table$Final |
                    Final.table$ComCol.recode,
type="percent",   # Note:  percent
par.settings=simpleTheme(lwd=2),
par.strip.text=list(cex=1.15, font=2),
scales=list(cex=1.15),
main="Histograms (Percent) of Final Exam Values by
Community College Graduation Status",
xlab=list("Final Exam Values (Limit = 0 to 100)",
cex=1.15, font=2),
xlim=c(0,120),
ylab=list("Percent", cex=1.15, font=2),
aspect=0.2,
layout = c(1,2),  # Note:  1 Column by 2 Rows.
col="red")
```

The layout argument can be used to place output into a variety of configurations, in a column by row format. Experiment with this argument.

The use of both count and percent breakout histograms from the `lattice::histogram()` function is clearly demonstrated in this set of teaching method and Community College figures. The count argument produced a set of histograms that may be somewhat difficult to interpret given the largely unequal N for each teaching method group. The histograms based on percent give a different perspective of distribution. Whether these figures are ever published, these distribution patterns should always be confirmed before the use of any inferential statistical tests.

The syntax for production of these many images is easily modified for the current study or for future studies. Once R syntax has been put into desired form, it is common practice to use and reuse this tested syntax for other applications.

2.4   Descriptive Analysis of the Data

Now that the data have been organized into the desired format and the `.recode` objects have also been created, a few more actions should be attempted before inferential tests (e.g., Student's T-test, ANOVA, MANOVA, etc.) are organized. Specifically, it is useful to carefully examine the dataset using summary and breakout descriptive statistics. Common graphical presentations are also helpful, to have a good understanding of the data and to see if the assumptions associated with most inferential tests, such as normal distribution, can be accepted.

2.4.1 Summary Descriptive Statistics

For this presentation, descriptive statistics will be prepared for the continuous object variable `Final.table$Final` and the factor object variables `Final.table $Method.recode` and `Final.Table$ComCol.recode`. A variety of R functions that support descriptive statistics at the summary level include: `mean()`, `median()`, `summary()`, `psych::describe()`, and `epicalc::summ()`. Review the level of detail for each function.

```
mean(Final.table$Final, na.rm=TRUE)
median(Final.table$Final, na.rm=TRUE)
summary(Final.table$Final)

par(ask=TRUE)
epicalc::summ(Final.table$Final, # Use the epicalc package.
  by=NULL,
  graph=TRUE,
  box=TRUE,        # Generate a boxplot.
  pch=18,
  ylab="auto",
  main="Sorted Dotplot and Boxplot of Final.table$Final",
  cex.X.axis=1.15, cex.Y.axis=1.15,
  font.lab=2, dot.col="auto")
```

Using the by=NULL argument, the `epicalc::summ()` function does not provide breakout statistics, but instead provides descriptive statistics and an accompanying graphic for all `Final.table$Final` values.

```
install.packages("psych")
library(psych)          # Load the psych package.
help(package=psych)     # Show the information page.
sessionInfo()           # Confirm all attached packages.

psych::describe(Final.table$Final)
```

The `psych::describe()` function can be applied against the entire dataset. The desired output can then be copied and pasted into a formal report. Merely delete the output applied against numbers that are instead factor-type object variables, which are marked with a * symbol.

```
psych::describe(Final.table)
```

The `epicalc::summ()` function provides output that is somewhat similar to `psych::describe()` when applied against the full set of data, organized in the data frame. As a simple quality assurance tool, be sure to always look at the min (Minimum) and max (Maximum) output, to see if there are data that are simply out-of-range (e.g., a 3, 7, etc., showing as min or max output when the scale only supports 1 and 2, as codes for a factor object variable).

```
epicalc::summ(Final.table)
```

2.4.2   Breakout Descriptive Statistics

Descriptive statistics at the summary level are essential, but breakout statistics are also needed to gain a more complete understanding of the data. There are many ways to obtain breakout statistics, but the tapply() function, the psych:: describe.by() function, the epicalc::summ() function, the doBy:: summaryBy(), and the prettyR::brkdn() function are among the most detailed and easiest to use, to discern differences between breakout groups such as the four breakout groups for the object variable Final.table $Method.recode associated with this sample lesson (lecture, CBT, video, and IDS) and the two breakout groups for the object variable Final.table $ComCol.recode associated with this sample lesson (Yes—is a CC Graduate, and No—not a CC Graduate).

```
# tapply() Function
tapply(Final, Method.recode, summary, na.rm=TRUE,
  data=Final.table)
tapply(Final, ComCol.recode, summary, na.rm=TRUE,
  data=Final.table)

# psych::describe.by() Function
psych::describe.by(Final.table$Final, Final.table$Method.recode)
psych::describe.by(Final.table$Final, Final.table$ComCol.recode)

# epicalc::summ() Function
par(ask=TRUE)
epicalc::summ(Final.table$Final,
  by=Final.table$Method.recode,  # Note the use of by.
  graph=TRUE, # Use graph=TRUE,
  pch=20,      # if desired.
  ylab="auto",
  main="Sorted Dotplot of Final.table$Final
  by Final.table$Method.recode",
  cex.X.axis=0.95, cex.Y.axis=0.95,
  font.lab=2, dot.col="auto")

# epicalc::summ() Function
par(ask=TRUE)
epicalc::summ(Final.table$Final,
  by=Final.table$ComCol.recode,
  graph=TRUE, # Use graph=TRUE,   # Note the use of by.
  pch=20,      # if desired.
  ylab="auto",
```

```
    main="Sorted Dotplot of Final.table$Final
    by Final.table$ComCol.recode",
    cex.X.axis=0.95, cex.Y.axis=0.95,
    font.lab=2, dot.col="auto")

# doBy::summaryBy() Function
install.packages("doBy")
library(doBy)              # Load the doBy package.
help(package=doBy)         # Show the information page.
sessionInfo()              # Confirm all attached packages.

doBy::summaryBy(Final ~ Method.recode +
                    ComCol.recode,
                    data=Final.table,
                    FUN=c(mean,sd),
                    na.rm=TRUE,
                    keep.names=TRUE,
                    order=TRUE)
```

It will be difficult to accommodate missing values for length. The enumerated function below takes care of this problem.

```
descriptivefun <- function(x, ...){
  c(m=mean(x, ...), sd=sd(x, ...), l=length(x))
}

doBy::summaryBy(Final ~ Method.recode +
                    ComCol.recode,
                    data=Final.table,
                    FUN=descriptivefun,
                    na.rm=TRUE,
                    keep.names=TRUE,
                    order=TRUE)

# prettyR::brkdn() Function
install.packages("prettyR")
library(prettyR)           # Load the prettyR package.
help(package=prettyR)      # Show the information page.
sessionInfo()              # Confirm all attached packages.

prettyR::brkdn(Final ~ Method.recode,
  data=Final.table,
  maxlevels=4,
  num.desc=c("mean", "sd", "valid.n"),
  width=25, round.n=2)

prettyR::brkdn(Final ~ ComCol.recode,
  data=Final.table,
  maxlevels=4,
  num.desc=c("mean", "sd", "valid.n"),
  width=25, round.n=2)
```

```
prettyR::brkdn(Final ~ (Method.recode +
                        ComCol.recode),
  data=Final.table,
  maxlevels=4,
  num.desc=c("mean","sd","valid.n"),
  width=25, round.n=2)

prettyR::brkdn(Final ~ (ComCol.recode +
                        Method.recode),
  data=Final.table,
  maxlevels=4,
  num.desc=c("mean","sd","valid.n"),
  width=25, round.n=2)
```

With the collapsed and breakout statistics completed, the data are now ready for further analyses, such as two-way ANOVA for this sample.

A listing and demonstration of other R functions related to descriptive statistics, either at the summary level or for breakout groups, could go on for pages. More examples are demonstrated in the remaining lessons.

2.5 Use R for Two-Way Analysis of Variance (ANOVA)

The data have now been brought into this R session, data were organized and labeled to accommodate human cognition, graphical images have been produced at the summary level and breakout levels, descriptive statistics were generated at the summary level and breakout levels, and data have also been organized into simple and complex contingency tables.

With all actions now in final form, the data are ready for inferential analyses, such as two-way ANOVA for this sample.

The preceding graphical images and descriptive statistics, both summary descriptive statistics and breakout descriptive statistics, provide a fairly good idea of the final exam values, overall and more importantly for the purpose of this two-way ANOVA sample by breakout groups (Method.recode and ComCol.recode).

The task now is to use two-way ANOVA, as supported in R by the many available packages and functions, to determine if there are statistically significant differences between (1) the four teaching method breakout groups, (2) the two Community College graduation status breakout groups, and (3) interaction between teaching method and Community College graduation status.

The descriptive statistics and graphical images serve as a basis for an initial suggestion that differences may or may not exist, but only an inferential test can provide the precise assessment needed for informed judgment.

R supports many possible ways to perform a two-way ANOVA. A few methods that support two-way ANOVA are detailed below.

2.5.1   Two-Way ANOVA: `aov()` Function

Use the formula for a two-way factorial design ANOVA, which is typically represented as:

```
fit1 <- aov(y ~ A + B + A:B, data=dataframe)
summary(fit1)

fit2 <- aov(y ~ A*B, data=dataframe)
summary(fit2)
```

```
y   = Measured datum, (e.g., weight, exam score, etc.)
A   = Factor Variable A (e.g., Gender, Race-Ethnicity, etc.)
B   = Factor Variable B (e.g., Soil Type, Breed Type, etc.)
A:B = Interaction of Factor Variable A and Factor Variable B
```

Both ANOVA formulas yield the same result, but the following formula (e.g., algorithm) is likely more illustrative:

```
fit1 <- aov(y ~ A + B + A:B, data=dataframe)

Sample1.fit1 <- aov(Final ~ Method.recode +
                           ComCol.recode +
                           Method.recode:ComCol.recode,
                data=Final.table)
summary(Sample1.fit1)
#                      Df Sum Sq Mean Sq F value Pr(>F)
# Method.recode         3 5372.3 1790.78 23.7512 1.388e-10 ***
# ComCol.recode         1  153.8  153.84  2.0404    0.1578
# Method.recode:
#   ComCol.recode       3  178.0   59.33  0.7869    0.5055
# Residuals            67 5051.6   75.40

Sample1.fit2 <- aov(Final ~ Method.recode*ComCol.recode,
                data=Final.table)
summary(Sample1.fit2)

#                      Df Sum Sq Mean Sq F value Pr(>F)
# Method.recode         3 5372.3 1790.78 23.7512 1.388e-10 ***
# ComCol.recode         1  153.8  153.84  2.0404    0.1578
# Method.recode:
#   ComCol.recode       3  178.0   59.33  0.7869    0.5055
# Residuals            67 5051.6   75.40
```

To gain a sense of the descriptive statistics (summary and breakout), use the `model.tables()` function for another view of Grand Mean, Mean, and N ("rep" for replication in this output) for each cell in the factorial table.

```
print(model.tables(Sample1.fit1,"means"), digits=3)
print(model.tables(Sample1.fit2,"means"), digits=3)
```

As a throwaway diagnostic, use the `plot.design()` function shown below to see general trends for final examination scores by each breakout group.

```
par(ask=TRUE)
plot.design(Final ~ Method.recode + ComCol.recode,
  data=Final.table,
  main="Final Exam Values by Teaching Method and
  Community College Graduation Status",
  lwd=3, font=2, cex.lab=1.50, cex.axis=1.50,
  font.main=2, font.lab=2, font.axis=2)
```

Then, use the `interaction.plot()` function for more details, but now focusing on the object variables of greatest interest.

```
par(ask=TRUE)
interaction.plot(Final.table$Method.recode,
                 Final.table$ComCol.recode,
                 Final.table$Final,
  main="Interaction Plot:   Instructional Method, Status as a
  Community College Graduate, and Final Examination Score",
  font.lab=2, col=2:9, lty="solid")
```

2.5.2   Outcome to Sample 1

The many R functions demonstrated in this sample have been used to "hammer" the data, to use an American expression. These many functions were used to demonstrate the important nuances of the data. It is a minimal set of analyses for a two-way ANOVA that only provides descriptive statistics, a simple plot or two, and the corresponding ANOVA output table with F-values and p-values. Data are too valuable and outcomes are too important to conduct only the minimal actions.

In conclusion for Sample 1, there are overall statistically significant difference ($p <= 0.05$) in final exam values by teaching method (view the three $***$ symbols in the two-way ANOVA table, indicating that p is actually far less than 0.05). There was no observed statistically significant difference ($p <= 0.05$) in final exam values by Community College graduation status. Equally, there was no observed statistically significant interaction ($p <= 0.05$) between teaching method and Community College graduation status.

The following output (gained by using copy and paste) from the doBy:: `summaryBy()` function provides a fairly good idea of where differences occur by teaching method.

```
# Method.recode Final.mean Final.sd
#       Lecture    76.500   12.2774
#           CBT    92.000    4.1675
#         Video    91.381    5.1037
#           IDS    73.000   11.4601
```

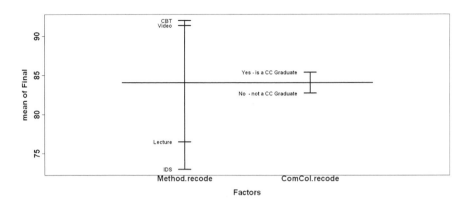

Fig. 2.1 Final exam values by Teaching Method and Community College Graduation status

However, to be precise, it is best to now apply the post hoc `TukeyHSD()` function, which is associated with one-way ANOVA, against final exam scores and teaching method (see Fig. 2.1).

```
Final.by.TeachingMethod.OnewayANOVA <-
   aov(Final ~ Method.recode, data=Final.table)

Final.by.TeachingMethod.OnewayANOVA

summary(Final.by.TeachingMethod.OnewayANOVA)

TukeyHSD(Final.by.TeachingMethod.OnewayANOVA)
   # Multiple comparisons.
   # This analysis accommodates missing data.

par(ask=TRUE)
plot(TukeyHSD(Final.by.TeachingMethod.OnewayANOVA),
   main="\n\nTukeyHSD Mean Comparison of Teaching Method
   and Final Exam Scores",
   cex.main=0.95, cex.lab=0.75, cex.axis=0.75,
   col.axis="darkblue", font.lab=2, font.axis=2, col="red")
   # Plot a graphical reinforcement of Mean Comparison
   # results, based on the TukeyHSD() function.
```

The output from use of the `TukeyHSD()` function provides the evidence needed to determine that there is a statistically significant difference ($p <= 0.05$) in final exam scores between: (1) lecture and CBT, (2) lecture and video, (3) IDS and CBT, and (4) IDS and video.

Conversely, there is no statistically significant difference ($p <= 0.05$) in final exam scores between: (1) IDS and lecture and (2) CBT and video.

```
#   Tukey multiple comparisons of means
#
```

```
# Fit: aov(formula = Final ~ Method.recode, data = Final.table)
#
# Method.recode
#                     diff        lwr       upr    p adj
# CBT-Lecture     15.50000     8.0569   22.9431  0.00000
# Video-Lecture   14.88095     7.5223   22.2396  0.00001
# IDS-Lecture     -3.50000   -11.3715    4.3715  0.64771
# Video-CBT       -0.61905    -7.7768    6.5387  0.99581
# IDS-CBT        -19.00000   -26.6840  -11.3160  0.00000
# IDS-Video      -18.38095   -25.9832  -10.7787  0.00000
```

Given these outcomes, it may be best to ignore any concern about prior status for students who are Community College graduates. Instead, focus should be given to teaching method. Assuming these outcomes stand the test of replication, efforts should be directed to teaching by use of CBT and video and efforts should be directed away from lecture and IDS.

Given the need for attention to policy considerations related to documentation of student learning outcomes, replication is a keyword in this broad conclusion. Outcomes across various scenarios, various faculty members, various courses, etc., would be needed before policy would ever be directed away from lecture as a desired teaching method. This sample lesson is merely one attempt in this overall assessment of student learning outcomes.

```
###############################################################
Prepare to Exit, Save, and Later Retrieve This R Session #
###############################################################

getwd()             # Identify the current working directory.
ls.str()            # List all objects, with finite detail.
save.image("Two-Way_ANOVA_Education-FinaExam.rdata")
list.files()        # List files at the PC directory.
q()                 # Quit this session.
                    # Prepare for Save workspace image? query.
####################### END ###########################
```

Use the R graphical user interface (GUI) to load the saved .rdata file: File and then load workspace. Otherwise, use the load() function, keying the full pathname, to load the .rdata file and retrieve the session. Recall, however, that it may be just as useful to simply use the .R script file and recreate the analyses and graphics, provided the data files remain available.

Chapter 3
Two-Way Analysis of Variance (ANOVA) Sample 2: Comparison of Systolic Blood Pressure Readings by Self-Declared Smoking Habits and by Self-Declared Drinking Habits

Abstract The complexity presented in this chapter increases to a measured degree, where R is used as the main tool for analyses of a wellness inventory. Detailed instructions are provided on how R is used to import data in a fixed width–fixed column file. Emphasis is placed on graphical figures as a visual means of conducting quality assurance, with new graphical tools added to this chapter. New packages, with more useful R functions than previously presented, are introduced in this chapter. Complexities from the real-world occurrence of missing data are addressed, both on the practical implications of missing data as well as how R is used to accommodate missing data.

The data for this set of analyses are the outcome of a special activity, where a beginning health professions intern queried a nonrandom sample (e.g., a convenience sample) of family members, friends, and other contacts and in turn obtained data on various items related to general health.

Some data were gained by using different instruments or measuring devices, dependent on the skill set of the intern and the level of supervision by the clinical professor: height in inches, weight in pounds, waist in inches, systolic blood pressure (mmHg), diastolic blood pressure (mmHg), resting heart rate, and total blood cholesterol.

Other data were patient responses to a wellness inventory, which was prepared by the intern after the benefit of literature review and approval by the supervising clinical professor and institutional review board: gender, date-of-birth (Month/Day/Year, XX/XX/XXXX), race, sleep hours last night, smoke, drink, income, exercise, and food from home.

After the process was completed and all data were obtained, each patient was assigned a patient identification number, ranging from 1 to 120, based on a last name ascending order alpha sort. This practice was judged necessary by the clinical professor in an attempt to maintain better command and control of the dataset.

T.W. MacFarland, *Two-Way Analysis of Variance: Statistical Tests and Graphics Using R*, SpringerBriefs in Statistics 1, DOI 10.1007/978-1-4614-2134-4_3,
© Thomas W. MacFarland 2012

There are missing data in the dataset. Perhaps the intern did not attempt to obtain a specific measure. Or, an attempt was made but the result, if obtained, was judged by the clinical professor to be unacceptable. Possibly the patient was either unable to participate or did not wish to participate in the full set of data collection activities.

When viewing the measured data, accuracy should always be questioned. Were the data collection protocols consistent? Consider resting heart rate. When was this datum obtained from individual patients? Was it obtained in the early morning, after a restful sleep? Or, was it obtained later, in the evening, after eight hours of work and a stressful commute in heavy traffic? Perhaps more importantly, was it obtained in a consistent manner and at a consistent time for each patient? There are also missing data for the self-declared variables, such as smoking habits and drinking habits. Some patients do not always respond to all questions.

Self-declared data also need to be questioned in terms of accuracy, with consideration of reliability and validity, from among those patients who did respond. Consider responses to the query about daily exercise. Does two hours of running, as training for a marathon race, equate to two hours of golf, where an electric golf cart is used to ride from one hole to another? Consistency (e.g., reliability) is always a concern.

Consider, also, patient truthfulness. To be kind, some patients may be fuzzy on the number of cigarettes they smoke in a typical day, or the number of alcoholic drinks they drink in a typical week. Self-declared data are useful, but they must also be suspect. Truthfulness (e.g., validity) is always a concern. Yet, absent constant monitoring, which is nearly impossible to accommodate and may introduce atypical behavior, health practitioners must only too often work with health inventory data that are self-declared and may not be as reliable and valid as desired.

This study examines if there are differences in systolic blood pressure readings by self-declared smoking habits (four breakout groups) and by self-declared drinking habits (four breakout groups), from among patients included in a health intern's convenience sample. The intern used a manual sphygmomanometer to obtain systolic blood pressure (SBP) readings, subject to supervision by the clinical professor.

Details for smoking habits breakout groups are: (1) never smoked, (2) no longer smokes, (3) 1–20 cigarettes a day, and (4) greater than 20 cigarettes a day. There is no distinction between someone who no longer smokes and who gave up smoking in the last three months and someone who no longer smokes and who gave up smoking more than ten years ago. There is also no differentiation in cigarette type (e.g., filtered or unfiltered), cigarette brand, or general approach to smoking (e.g., someone who plays with a cigarette as compared to someone who draws heavily on a cigarette).

Details for drinking habits (Alcohol) breakout groups are: (1) never drank, (2) no longer drinks, (3) 1–5 drinks a week, and (4) greater than 5 drinks a week. There is no distinction between someone who no longer drinks and who gave up the drink in the last three months and someone who no longer drinks and who gave up the drink more than ten years ago. There is also no differentiation in selected drinks (e.g., light beer, 3.2 beer, regular beer, bourbon, gin, whiskey, etc.), brand, or general approach to drinking (e.g., someone who drinks only a small amount regularly throughout the week as compared to a binge drinker who rarely drinks but then drinks in excess when drink is finally taken).

Along with reliability and validity, the issue of scale or measurement must be considered. Because the clinical professor, in the role of supervising professional, was confident that systolic blood pressure readings were accurate, it was judged that this measure represented interval data. There are concerns, however, about the scale for the two factor object variables, smoke and drink. The scale for these two factor object variables is clearly not interval, but is the scale ordinal? That is to say, does this scale represent an ordering of any type? Or, are the selections for smoke and drink best viewed from the perspective of nominal (e.g., named) data, only? For this sample lesson, smoke and drink will be viewed as nominal data, to be overly conservative on this issue. But of course, there are those who could argue the case that the data for smoke and drink are ordinal.

Given this background, two-way analysis of variance (ANOVA) was judged to be the appropriate test for factorial-type analysis of summative differences in systolic blood pressure readings by smoking habits (four breakout groups) and by drinking habits (four breakout groups). The first three lines of data and the last three lines of data, presented in fixed width–fixed column (e.g., space-separated) format, are shown below (note how there is no header to this file):

```
  1 F 07/24/1972  51 A 221 42 130   80   80 221  5 2 4 1 0 1
  2 F 09/01/1977  63 W 112 30  96   62   74 167  6 2 2 2 1
  3 M 06/26/1967  69 H 148 30 124   68   72 170  9 1 3 4 2 1
118 M 03/04/1975  65 W 169 32 122   78   64 179  5 4 3 5 0 2
119 M 01/10/1976  66 W 175 32 124   84   60 190  6 2 2 3 2 1
120 M 02/27/1961  69 W 210 43 152   86   82 231  7 4 4 2 1 3
```

Sample 2 Ho (Null Hypothesis): There is no difference between smoking habits, drinking habits, and interaction between smoking habits and drinking habits regarding systolic blood pressure readings from among patients in a convenience sample ($p <= 0.05$).

3.1   Data Import of a .prn Fixed Width–Fixed Column (e.g., Space-Separated) Format Data File into R

The dataset for this lesson is somewhat different from what is now likely the norm, purposely as an interesting and too often overlooked demonstration. For this lesson, notice how the ASCII file represents a dataset that has been prepared in .prn fixed width–fixed column format (not .csv format) and that the data are aligned in strict width–column placement. Neither commas nor tabs were used to separate the data. Instead, spaces are used to separate the data. Experienced researchers will work with data in many formats, so it is desirable to gain experience with this type of file format even though .csv (comma-separated values) data files are certainly quite common when using R and other data analysis tools.

For this sample, assume that the data were initially entered into a commercial spreadsheet and that the data were then saved in .prn file format by using File - > Save As - > Formatted Text (Space Delimited). The data were then imported into an open source text editor (vim is merely one of many possible selections for an open source text editor) for review and editing, as needed. Assume that the .prn file extension was retained when the dataset was put into final form. It is common for data to go through extensive review processes and an experienced researcher will know how to work with data in multiple formats. The desire here is to maintain quality assurance while considering the use of low cost or no cost software that is still of the highest quality.

Because the data are in fixed width–fixed column format, with spaces separating data, it should be a fairly easy action to import the data into R using the read.fwf() function. Data files in .csv file format are quite common, but if care and attention are given to syntax and column placement then it is also quite easy to import fixed width–fixed column format files into R. The data for this lesson, in fixed width–fixed column (e.g., space-separated) format as an ASCII .prn file, are found at F:\R_Lessons\Inferential_Statistics_Parametric on a standalone personal computer.

All analyses begin from this starting point, working with a previously prepared fixed width–fixed column format .prn ASCII file. The emphasis will be on: (1) overall analysis of systolic blood pressure, (2) breakout analyses of systolic blood pressure: systolic blood pressure by cigarette smoking habits and systolic blood pressure by alcohol drinking habits, (3) graphical representation of overall and breakout findings, (4) two-way ANOVA of the data, and (5) summative interpretation of outcomes.

Note below how the read.fwf() function is used to read in the fixed width–fixed column (e.g., fwf—fixed width format) format ASCII-text file, which has been saved with a .prn file extension. The .prn file could have just as easily been saved in ASCII format with either a .txt or .dat extension. The .prn file extension was merely selected to show another example of how R can accommodate multiple file formats and multiple file extensions.

```
###########################################################
# Housekeeping            Use for All Analyses
###########################################################
setwd("F:/R_Lessons/Inferential_Statistics_Parametric")
                    # Set to a new working directory.
                    # Note the single forward slash and double
                    # quotes.
                    # This new directory should be the directory
                    # where the data file is located, otherwise
                    # the data file will not be found.
getwd()             # Confirm the working directory.
search()            # Attached packages and objects.
###########################################################
```

3.1.1   Data Import or Data Entry

```
Health.table <- read.fwf("Wellness-Inventory_fixed-columns.prn",
  header=FALSE,   # There is no header for column names.
  skip=0,         # Skip no lines; read data from the first line.
  na.strings=" ",# Blank spaces in a character object are NA.
  width=c(-1,3,   # Patient Identification Number
          -1,1,   # Gender
          -1,10,  # Date-of-Birth (Month/Day/Year, XX/XX/XXXX)
          -1,3,   # Height in Inches
          -1,1,   # Race
          -1,3,   # Weight in Pounds
          -1,2,   # Waist in Inches
          -1,3,   # Systolic Blood Pressure (mmHg)
          -1,3,   # Diastolic Blood Pressure (mmHg)
          -1,3,   # Resting Heart Rate
          -1,3,   # Total Blood Cholesterol
          -1,2,   # Sleep Hours Last Night
          -1,1,   # Smoke
          -1,1,   # Drink
          -1,1,   # Income
          -1,1,   # Exercise
          -1,1))  # Food From Home
names(Health.table) <- c(
  "ID",           # Patient Identification Number
  "Gender",       # Gender
  "DOB",          # Date-of-Birth (Month/Day/Year, XX/XX/XXXX)
  "Height",       # Height in Inches
  "Race",         # Race
  "Weight",       # Weight in Pounds
  "Waist",        # Waist in Inches
  "SBP",          # Systolic Blood Pressure (mmHg)
  "DBP",          # Diastolic Blood Pressure (mmHg)
  "RHR",          # Resting Heart Rate
  "TBC",          # Total Blood Cholesterol
  "Sleep",        # Sleep Hours Last Night
  "Smoke",        # Smoke
  "Drink",        # Drink
  "Income",       # Income
  "Exercise",     # Exercise
  "Food")         # Food From Home

getwd()                  # Identify the working directory.
ls()                     # List objects.
attach(Health.table)     # Attach the data, for later use.
names(Health.table)      # Identify names.
head(Health.table)       # Show the head.
tail(Health.table)       # Show the tail.
Health.table             # Show the entire data frame.
```

As the names() function is used, the names are declared in the sequential order
as the object variables are presented, moving from the first (ID) to the last (Food).
As a hint, it is best to avoid blank spaces when declaring a name. Use a meaningful
string and then embellish the name later, using some type of label accommodation.

Now that the fixed width–fixed column format file dataset has been imported
into R, it is perhaps necessary to deconstruct the meaning of the different
lines of syntax shown above, such as, $-1, 3$ to $-1, 1$ given how this is a
different approach to identification of column placement. The important thing
to remember is that the data are indeed in fixed width–fixed column format, with
data in specific columns. As an example, look at the first five lines (Patients
1–5) of data and the last five lines (Patients 116–120) of data in the file
Wellness-Inventory_fixed-columns.prn. A self-constructed column
placement ruler has been placed immediately before and immediately after the
data (shown below), to assist with understanding column ordering of the 17 object
variables in this dataset.

```
Column Placement Ruler
          1         2         3         4         5         6
123456789012345678901234567890123456789012345678901234567890

  1 F 07/24/1972  51 A 221 42 130  80  80 221  5 2 4 1 0 1
  2 F 09/01/1977  63 W 112 30  96  62  74 167  6 2 2 2 1
  3 M 06/26/1967  69 H 148 30 124  68  72 170  9 1 3 4 2 1
  4 M 10/08/1979  72 W 130 30 110  70  69 162  4 4 4 5 3 3
  5 F 10/18/1972  74    210 40 140  82  83 218  8 4 2 5 1 2

116 F 07/24/1964  70 B 190 38 160  90 100 174  8 3 1 1 2
117 F 01/01/1982  68 A 164 30 110  60  90 168  9 2   1 1
118 M 03/04/1975  65 W 169 32 122  78  64 179  5 4 3 5 0 2
119 M 01/10/1976  66 W 175 32 124  84  60 190  6 2 2 3 2 1
120 M 02/27/1961  69 W 210 43 152  86  82 231  7 4 4 2 1 3

123456789012345678901234567890123456789012345678901234567890
          1         2         3         4         5         6
Column Placement Ruler
```

Note how the first column is blank and the next three columns hold the data
for patient (e.g., 1–120). The width argument accommodates this column structure
by skipping one column (using -1) and then reading in data from the next three
columns (using 0.3). There is then one blank column (using -1) and the data for
gender are then found in the next singular column (using 0.1). Then, notice how
there is one blank column (using -1) followed by Date-of-Birth data in the next
ten columns (using 0.10). This structure for the dataset continues until the column
placement for the last object variable (Food From Home) is reached and marked
as $-1, 1$.

This arrangement of identifying blank columns and then filled columns takes
some practice, but fixed width–fixed column format files are still fairly common

(especially for large datasets from some government resources) and experienced researchers need to work with data in multiple formats. (Fixed width–fixed column format files also negate possible problems from an errant comma, which may occur when working with a .csv (comma-separated values) file.) We do not always have the luxury of selecting file formats when data are provided by third party resources.

As a final comment on working with fixed width–fixed column format files, be sure to accommodate extreme values when declaring column placement. Consider how: (1) patient identification number requires three columns, to accommodate patients 100–120, (2) sleep requires two columns, to account for 10, 11, 12, etc., hours of sleep, and (3) height in inches needs three columns, to allow for the extreme case of anyone with a height of 100 in (8 ft 4 in) or more. Know your data and anticipate the unexpected.

By completing these many actions, an object called Health.table has been created. This object consists of the data included in the fixed width–fixed column file. Make sure that there are no prior R-based datasets called Health.table available. Note how it was only necessary to key the filename for the .prn file and not the full pathname since the R working directory is currently set to the directory and subdirectory where this .prn file is located (see Sect. 3.1 at the beginning of this lesson).

3.2   Organize the Data and Display the Code Book

Now that the data have been imported into R, it is usually necessary to check the data for format and then make any changes that may be needed, to organize the data. This concern is especially important for numeric codes used to identify factors (e.g., groups) for the object variables smoke, drink, income, exercise, and food. Prior to the use of these object variables, which all have numeric codes, recall how the Gender object variable (e.g., *F* and *M*) and the race object variable (e.g., *A, B, H, W, X*) used alpha characters to identify various factors (e.g., groups).

```
class(Health.table)
class(Health.table$ID)        # DataFrame$ObjectName notation.
class(Health.table$Gender)    # DataFrame$ObjectName notation.
class(Health.table$DOB)       # DataFrame$ObjectName notation.
class(Health.table$Height)    # DataFrame$ObjectName notation.
class(Health.table$Race)      # DataFrame$ObjectName notation.
class(Health.table$Weight)    # DataFrame$ObjectName notation.
class(Health.table$Waist)     # DataFrame$ObjectName notation.
class(Health.table$SBP)       # DataFrame$ObjectName notation.
class(Health.table$DBP)       # DataFrame$ObjectName notation.
class(Health.table$RHR)       # DataFrame$ObjectName notation.
class(Health.table$TBC)       # DataFrame$ObjectName notation.
class(Health.table$Sleep)     # DataFrame$ObjectName notation.
class(Health.table$Smoke)     # DataFrame$ObjectName notation.
```

```
class(Health.table$Drink)     # DataFrame$ObjectName notation.
class(Health.table$Income)    # DataFrame$ObjectName notation.
class(Health.table$Exercise)  # DataFrame$ObjectName notation.
class(Health.table$Food)      # DataFrame$ObjectName notation.
```

Now that the class() function has been applied against each object, consult
the code book and coerce each object, as needed, into its correct class. Typically,
integers that serve as numeric codes (e.g., 1 represents blue and 2 represents green)
are coerced into factor format.

```
# Code Book ###############################################
##########################################################
# Wellness Inventory Code Book
#
# Variable Labels
#   ID          Patient Identification Number
#   Gender      Gender
#   DOB         Date-of-Birth (Month/Day/Year, XX/XX/XXXX)
#   Height      Height in Inches
#   Race        Race
#   Weight      Weight in Pounds
#   Waist       Waist in Inches
#   SBP         Systolic Blood Pressure (mmHg)
#   DBP         Diastolic Blood Pressure (mmHg)
#   RHR         Resting Heart Rate
#   TBC         Total Blood Cholesterol
#   Sleep       Sleep Hours Last Night
#   Smoke       Smoke
#   Drink       Drink
#   Income      Income
#   Exercise    Exercise
#   Food        Food From Home
#
# Variable Values
#   ID          Nominal   LOW to HIGH
#   Gender      Nominal   F Female
#                         M Male
#   DOB         Date      Month/Day/Year (XX/XX/XXXX)
#   Height      Interval  LOW to HIGH
#   Race        Nominal   A Asian
#                         B Black
#                         H Hispanic
#                         W White
#                         X American Indian
#   Weight      Interval  LOW to HIGH
#   Waist       Interval  LOW to HIGH
#   SBP         Interval  LOW to HIGH
#   DBP         Interval  LOW to HIGH
#   RHR         Interval  LOW to HIGH
#   TBC         Interval  LOW to HIGH
#   Sleep       Interval  LOW to HIGH
```

```
#    Smoke         Nominal     1 Never smoked
#                              2 No longer smokes
#                              3 1 to 20 Cigarettes a day
#                              4 > 20 Cigarettes a day
#    Drink         Nominal     1 Never drank
#                              2 No longer drinks
#                              3 1 to 5 Drinks a week
#                              4 > 5 Drinks a week
#    Income        Nominal     1 $00,000 to $24,999 a year
#                              2 $25,000 to $49,999 a year
#                              3 $50,000 to $74,999 a year
#                              4 $75,000 to $99,999 a year
#                              5 $100,000 or more a year
#    Exercise      Nominal     0 No Exercise
#                              1 1 to 149 minutes a week
#                              2 150 to 299 minutes a week
#                              3 300 minutes or more a week
#    Food          Nominal     0 0 Meals a week away from home
#                              1 1 to 5 Meals a week away from home
#                              2 6 to 10 Meals a week away from home
#                              3 > 10 Meals a week away from home
##############################################################
```

```
install.packages("epicalc")
library(epicalc)              # Load the epicalc package.
help(package=epicalc)         # Show the information page.
sessionInfo()                 # Confirm all attached packages.
```

In an effort to promote self-documentation and readability, it is often desirable to label all object variables. The epicalc label.var() function can serve this purpose. Load the epicalc package, if it is not operational from prior analyses.

```
epicalc::des(Health.table)
```

Use the epicalc::des() function to see the nature of the data frame. Then, provide a useful description of each object variable by using the epicalc::label.var() function. For this lesson, only those object variables in Health.table of direct importance to the two-way ANOVA will be accommodated here and throughout this lesson: SBP, smoke, and drink.

```
epicalc::label.var(SBP,        "Systolic Blood Pressure (mmHg)",
  dataFrame=Health.table)
epicalc::label.var(Smoke,      "Smoke",
  dataFrame=Health.table)
epicalc::label.var(Drink,      "Drink",
  dataFrame=Health.table)

epicalc::des(Health.table)
  # Confirm the description of each object variable.
```

Coerce objects into correct format. Notice how variables are named: `DataFrame$ObjectName`. At first this action may seem too formal and wordy, but it is actually very useful to ensure that actions are performed against the correct object. Most text editors allow the use of copy/paste and find/replace, so it should be a simple operation to organize the syntax.

```
# Object class before coercion
class(Health.table)
class(Health.table$ID)          # DataFrame$ObjectName notation.
class(Health.table$Gender)      # DataFrame$ObjectName notation.
class(Health.table$DOB)         # DataFrame$ObjectName notation.
class(Health.table$Height)      # DataFrame$ObjectName notation.
class(Health.table$Race)        # DataFrame$ObjectName notation.
class(Health.table$Weight)      # DataFrame$ObjectName notation.
class(Health.table$Waist)       # DataFrame$ObjectName notation.
class(Health.table$SBP)         # DataFrame$ObjectName notation.
class(Health.table$DBP)         # DataFrame$ObjectName notation.
class(Health.table$RHR)         # DataFrame$ObjectName notation.
class(Health.table$TBC)         # DataFrame$ObjectName notation.
class(Health.table$Sleep)       # DataFrame$ObjectName notation.
class(Health.table$Smoke)       # DataFrame$ObjectName notation.
class(Health.table$Drink)       # DataFrame$ObjectName notation.
class(Health.table$Income)      # DataFrame$ObjectName notation.
class(Health.table$Exercise)    # DataFrame$ObjectName notation.
class(Health.table$Food)        # DataFrame$ObjectName notation.

# Coercion
Health.table$ID        <- as.factor(Health.table$ID)
Health.table$Gender    <- as.factor(Health.table$Gender)
Health.table$DOB       <- as.Date(Health.table$DOB, "%m/%d/%Y")
Health.table$Height    <- as.numeric(Health.table$Height)
Health.table$Race      <- as.factor(Health.table$Race)
Health.table$Weight    <- as.numeric(Health.table$Weight)
Health.table$Waist     <- as.numeric(Health.table$Waist)
Health.table$SBP       <- as.numeric(Health.table$SBP)
Health.table$DBP       <- as.numeric(Health.table$DBP)
Health.table$RHR       <- as.numeric(Health.table$RHR)
Health.table$TBC       <- as.numeric(Health.table$TBC)
Health.table$Sleep     <- as.numeric(Health.table$Sleep)
Health.table$Smoke     <- as.factor(Health.table$Smoke)
Health.table$Drink     <- as.factor(Health.table$Drink)
Health.table$Income    <- as.factor(Health.table$Income)
Health.table$Exercise  <- as.factor(Health.table$Exercise)
Health.table$Food      <- as.factor(Health.table$Food)
```

Although it is not the primary purpose of this lesson, note how arguments were used to coerce `Health.table$DOB` into a fairly common date format: MM/DD/YYYY (e.g., 12/24/1989). Other date options are available with R, based on selected arguments and local preferences for the way dates are presented. Key help(as.Date) for more details on the way R accommodates dates.

```
# Object class after coercion
class(Health.table)
class(Health.table$ID)          # DataFrame$ObjectName notation.
class(Health.table$Gender)      # DataFrame$ObjectName notation.
class(Health.table$DOB)         # DataFrame$ObjectName notation.
class(Health.table$Height)      # DataFrame$ObjectName notation.
class(Health.table$Race)        # DataFrame$ObjectName notation.
class(Health.table$Weight)      # DataFrame$ObjectName notation.
class(Health.table$Waist)       # DataFrame$ObjectName notation.
class(Health.table$SBP)         # DataFrame$ObjectName notation.
class(Health.table$DBP)         # DataFrame$ObjectName notation.
class(Health.table$RHR)         # DataFrame$ObjectName notation.
class(Health.table$TBC)         # DataFrame$ObjectName notation.
class(Health.table$Sleep)       # DataFrame$ObjectName notation.
class(Health.table$Smoke)       # DataFrame$ObjectName notation.
class(Health.table$Drink)       # DataFrame$ObjectName notation.
class(Health.table$Income)      # DataFrame$ObjectName notation.
class(Health.table$Exercise)    # DataFrame$ObjectName notation.
class(Health.table$Food)        # DataFrame$ObjectName notation.
```

The epicalc package has many useful functions. Saying this, use the epicalc::des() function to again describe the data frame currently in use.

```
epicalc::des(Health.table)
```

Use the summary() function against the object Health.table, which is a data frame, to gain an initial sense of descriptive statistics and frequency distributions.

```
summary(Health.table)
```

Although the dataset seems to be in correct format, it is somewhat difficult to work with numeric values for the factor object variables in question for this sample lesson: smoke and drink. There are more than a few ways to apply labels to factor object variables when using R. The method used in this dataset is merely one example. Again, the focus here is on the factor object variables of interest to this two-way ANOVA. Use the code book to review the meaning for each factor code and then note how this problem is easy to accommodate.

```
# Apply the labels() function
Health.table$Smoke.recode  <- factor(Health.table$Smoke,
   labels = c("Never smoked",
              "No longer smokes",
              "1 to 20 Cigarettes a day",
              "> 20 Cigarettes a day"))

head(Health.table$Smoke) # View initial data.
head(Health.table$Smoke.recode)
```

```
summary(Health.table$Smoke) # View descriptive statistics.
summary(Health.table$Smoke.recode)

par(ask=TRUE)
epicalc::tab1(Health.table$Smoke.recode,
  decimal=2,                        # Use the tab1() function
  sort.group=FALSE,                 # from the epicalc
  cum.percent=TRUE,                 # package to see details
  graph=TRUE,                       # about the selected
  missing=TRUE,                     # object variable. (The
  bar.values=c("frequency"),        # 1 of tab1 is the one
  horiz=FALSE,                      # numeric character and
  cex=1.15,                         # it is not the letter
  cex.names=1.15,              .    # l).
  cex.lab=1.15,
  cex.axis=1.15,
  main="Factor Levels for Object Variable Smoke. recode",
  ylab="Frequency of Smoke.recode Factor Levels",
  col = c(rainbow(5)),
  gen=TRUE)
  # The number of colors in the rainbow() function was set
  # at one more than the number of levels for the factor
  # object variable, to better accommodate graphical
  # representation of NAs.

# Apply the labels() function
Health.table$Drink.recode  <- factor(Health.table$Drink,
  labels = c("Never drank",
             "No longer drinks",
             "1 to 5 Drinks a week",
             "> 5 Drinks a week"))

head(Health.table$Drink) # View initial data.
head(Health.table$Drink.recode)

summary(Health.table$Drink) # View descriptive statistics.
summary(Health.table$Drink.recode)

par(ask=TRUE)
epicalc::tab1(Health.table$Drink.recode,
  decimal=2,                        # Use the tab1() function
  sort.group=FALSE,                 # from the epicalc
  cum.percent=TRUE,                 # package to see details
  graph=TRUE,                       # about the selected
  missing=TRUE,                     # object variable. The
  bar.values=c("frequency"),        # 1 of tab1 is the one
  horiz=FALSE,                      # numeric character and
  cex=1.15,                         # it is not the letter
```

```
cex.names=1.15,                    # 1).
cex.lab=1.15,
cex.axis=1.15,

main="Factor Levels for Object Variable Drink.recode",
ylab="Frequency of Drink.recode Factor Levels",
col = c(rainbow(5)),
gen=TRUE)
# The number of colors in the rainbow() function was set
# at one more than the number of levels for the factor
# object variable, to better accommodate graphical
# representation of NAs.
```

To recap, the object variable Health.table$Smoke.recode was created by applying the factor() function against the object variable Health.table$Smoke. The labels() function was used to embellish future output into simple English. This approach to labeling the four smoke breakout groups and then confirming this action was then applied to the object variable drink, using the appropriate labels for drink. If needed, this action could be applied to all other object variables in this dataset. The head() function was used to view the first few lines of data. The summary() function was used to gain an initial sense of the data for the object variable in question.

Now, merely use the attach() function again to confirm that all data are attached to the data frame.

```
attach(Health.table)
head(Health.table)
tail(Health.table)
summary(Health.table)     # Quality assurance data check.

str(Health.table)         # List all objects, with finite detail.
```

As an additional data check, use the table() function to see how data have been summarized using the newly created names (factor object variables) as well as the original names for the numeric object variables.

```
table(Health.table$Smoke,          useNA = c("always"))
table(Health.table$Smoke.recode, useNA = c("always"))

table(Health.table$Drink,          useNA = c("always"))
table(Health.table$Drink.recode, useNA = c("always"))

table(Health.table$Smoke,
      Health.table$Drink, useNA = c("always"))

table(Health.table$Smoke.recode,
      Health.table$Drink.recode, useNA = c("always"))
```

Note how the argument useNA = c("always") is used with the table function, to force identification of missing values.

This type of redundancy and attention to detail at this stage of development may seem unnecessary, but it more than helps reduce later errors caused by a simple oversight. It takes less time to put data into correct format at the beginning of a project than the amount of time needed to make corrections midpoint or at the end of a project, if indeed corrections are possible. Quality assurance is a continuous process, from preplanning to end-of-project debriefing.

3.3 Conduct a Visual Data Check

The summary() function, min() function, and max() function are all certainly useful for data checking, but there are also many advantages to a visual data check process. In this case, simple plots and barcharts can be very helpful in an attempt to look for data that may be either illogical or out-of-range. These initial plots and barcharts will be, by design, simple, and should be considered throwaways as they are intended only for initial diagnostic purposes. They will then be followed by graphical images that provide more detail and may have future use in any possible presentation(s). Again, for this sample lesson the focus will be on SBP, smoke (including Smoke.recode), and drink (including Drink.recode).

```
names(Health.table)    # Confirm all object variables.
```

3.3.1 Simple Plots

```
par(ask=TRUE)
plot(Health.table$SBP,
  main="Health.table$SBP Visual Data Check",
  pch=19,
  col = c("black"))
  # Note how the datapoints are represented as
  # black solid circles for this numeric object
  # variable.

par(ask=TRUE)
plot(Health.table$Smoke,
  main="Health.table$Smoke Visual Data Check",
  col = c(rainbow(4)))
  # Factor object variable breakout groups.
```

```
par(ask=TRUE)
plot(Health.table$Smoke.recode,
  main="Health.table$Smoke.recode Visual Data Check",
  col = c(rainbow(4)))
  # Factor object variable breakout groups.

par(ask=TRUE)
plot(Health.table$Drink,
  main="Health.table$Drink Visual Data Check",
  col = c(rainbow(4)))
  # Factor object variable breakout groups.

par(ask=TRUE)
plot(Health.table$Drink.recode,
  main="Health.table$Drink.recode Visual Data Check",
  col = c(rainbow(4)))
  # Factor object variable breakout groups.
```

These simple plots offer a first view of the data and they also provide a sense of any outliers, if indeed there are data with extreme values. In an attempt to look for outliers, the ylim argument has been avoided, so that all data are plotted. This action will help with the identification of outliers, if present.

3.3.2   Histogram of the Summary Object Variable

This sample lesson has been designed to look into the nature of object variable SBP (systolic blood pressure) and the factor variables smoke (self-declared smoking habits) and drink (self-declared drinking habits), recoded into a more verbose format as object variables Smoke.recode and Drink.recode. Given the nature of SBP readings, it may also be a good idea to supplement the plot() function with other functions, to gain a different view of the continuous values of SBP, overall and by breakout groups.

First, consider the shape and distribution of a histogram under the ideal conditions of normal distribution. Use R in interactive mode to create a dataset with 100,000 datapoints. To make this sample interesting, assume that Mean = 50 and SD = 2.

```
SampleDataset <- rnorm(100000, 50, 2)

par(ask=TRUE)
hist(SampleDataset)
```

Then, compare the ideal SampleDataset histogram to the distribution of real-world values for Health.table$SBP.

```
par(ask=TRUE)
hist(Health.table$SBP,
  main="Health.table$SBP Visual Data Check (Histogram)",
  font=2, cex.lab=1.15, col="red")
```

Below, the hist() function is applied again against the object variable
Health.table$SBP but with more detail, by use of additional arguments
associated with the hist() function.

```
par(ask=TRUE)
hist(Health.table$SBP,
  main="Histogram of SBP Readings",
  xlab="SBP Readings (Limit = 0 to 200)",
  ylab="Frequency",
  xlim=c(0,220),      # Note the selection for xlim.
  ylim=c(0,30),       # Note the selection for ylim.
  cex.lab=1.15, cex.axis=1.15, freq=TRUE,
  border="blue", col="red",)
  # Again, note the xlim=c(0,220) argument.  By
  # design, the X axis is pushed out to 220
  # instead of 180, to accommodate a full
  # presentation of output.

# Create the object variable SBP.mean
SBP.mean <- mean(Health.table$SBP, na.rm=TRUE)
  # Note the need for na.rm=TRUE.
SBP.mean

# Create the object variable SBP.sd
SBP.sd   <- sd(Health.table$SBP, na.rm=TRUE)
  # Note the need for na.rm=TRUE.
SBP.sd


par(ask=TRUE)
hist(Health.table$SBP,
  main="Histogram of SBP Readings",
  xlab="SBP Readings (Limit = 0 to 220)",
  ylab="Representation",
  xlim=c(0,220),      # Note the selection for xlim.
  ylim=c(0,0.03),     # Note the selection for ylim.
  font.lab=2,
  font.axis=2,
  border="blue",
  col="red",
  nclass=15,
```

```
   prob=TRUE)
curve(dnorm(x, mean=SBP.mean, sd=SBP.sd),
   col="darkblue",
   lwd=2,
   add=TRUE)
box()
```

Notice how `xlim=c(0,220)` and `ylim=c(0,0.03)` were used to set the limits for the *X*-axis and the *Y*-axis. The *X*-axis was purposely set to go to 220 even though the maximum value of SBP is 180. This action extended the normal curve overlay to the right and allowed presentation of the entire curve. In the same manner, the *Y*-axis was set to allow presentation of the upper limits of the histogram. Experiment with these arguments, to see best placement in support of individual needs.

3.3.3   Horizontal and Vertical Boxplots of the Summary Object Variable

The boxplot is commonly seen in the literature, as a means of representing distribution of values. Depending on preference and need, a boxplot can be displayed in either horizontal or vertical orientation while still highlighting the lower quartile (Q1), the median (Q2), the upper quartile (Q3), and any possible outliers.

```
par(ask=TRUE)
boxplot(Health.table$SBP,
   horizontal=TRUE,
   main="Horizontal Boxplot of SBP Readings",
   xlab="SBP Readings (Limit = 0 to 220)",
   ylim=c(0,220),      # Note the selection for ylim.
   cex.lab=1.15,
   cex.axis=1.15,
   lty=1,              # Note the line type.
   lwd=3,              # Note the line width.
   border="blue",
   col="red")
box()

par(ask=TRUE)
boxplot(Health.table$SBP,
   horizontal=FALSE,
   main="Vertical Boxplot of SBP Readings",
   ylab="SBP Exam Values (Limit = 0 to 220)",
   ylim=c(0,220),      # Note the selection for ylim.
   cex.lab=1.15,
```

```
cex.axis=1.15,
lty=1,                 # Note the line type.
lwd=3,                 # Note the line width.
border="blue",
col="red")
box()
```

3.3.4   Histogram of the Summary Object Variable
by Breakout Object Variables

Experiment with the lattice package since it can provide informative graphics in a
fairly compact presentation, yet still of publishable quality.

3.3.4.1 Breakout Histograms by Count

```
install.packages("lattice")
library(lattice)        # Load the lattice package.
help(package=lattice)   # Show the information page.
sessionInfo()           # Confirm all attached packages.

par(ask=TRUE) # Smoke.recode 2 Columns by 2 Rows
lattice::histogram(~Health.table$SBP |
                   Health.table$Smoke.recode,
  type="count",        # Note:  count
  par.settings=simpleTheme(lwd=2),
  par.strip.text=list(cex=1.15, font=2),
  scales=list(cex=1.15),
  main="Histograms (Count) of SBP Readings
  by Smoking Habits",
  xlab=list("SBP Readings (Limit = 0 to 220)",
  cex=1.15, font=2),
  xlim=c(0,220),
  ylab=list("Count", cex=1.15, font=2),
  aspect=0.5,          # Note the aspect value.
  layout = c(2,2),     # Note:  2 Columns by 2 Rows.
  col="red")

par(ask=TRUE) # Drink.recode 1 Column by 4 Rows
lattice::histogram(~Health.table$SBP |
                   Health.table$Drink.recode,
  type="count",        # Note:  count
  par.settings=simpleTheme(lwd=2),
```

```
    par.strip.text=list(cex=1.15, font=2),
    scales=list(cex=1.15),
    main="Histograms (Count) of SBP Readings by
    Drinking Habits",
    xlab=list("SBP  Values (Limit = 0 to 220)",
    cex=1.15, font=2),
    xlim=c(0,220),
    ylab=list("Count", cex=1.15, font=2),
    aspect=0.2,
    layout = c(1,4),  # Note:  1 Column by 4 Rows.
    col="red")

par(ask=TRUE) # Smoke.recode  1 Column by 4 Rows
lattice::histogram(~Health.table$SBP |
                    Health.table$Smoke.recode,
    type="percent",   # Note:  percent
    par.settings=simpleTheme(lwd=2),
    par.strip.text=list(cex=1.15, font=2),
    scales=list(cex=1.15),
    main="Histograms (Percent) of SBP Readings by
    Smoking Habits",
    xlab=list("SBP Readings (Limit = 0 to 220)",
    cex=1.15, font=2),
    xlim=c(0,220),
    ylab=list("Percent", cex=1.15, font=2),
    aspect=0.2,
    layout = c(1,4),  # Note:  1 Column by 4 Rows.
    col="red")

par(ask=TRUE) # Drink.recode 2 Columns by 2 Rows
lattice::histogram(~Health.table$SBP |
                    Health.table$Drink.recode,
    type="percent",   # Note:  percent
    par.settings=simpleTheme(lwd=2),
    par.strip.text=list(cex=1.15, font=2),
    scales=list(cex=1.15),
    main="Histograms (Percent) of SBP Readings by
    Drinking Habits",
    xlab=list("SBP Readings (Limit = 0 to 220)",
    cex=1.15, font=2),
    xlim=c(0,220),
    ylab=list("Percent", cex=1.15, font=2),
    aspect=0.5,         # Note the aspect value.
    layout = c(2,2),  # Note:  2 Columns by 2 Rows.
    col="red")
```

The use of both count and percent breakout histograms from the `lattice::histogram()` function is clearly demonstrated in this set of smoking habits and drinking habits figures. The `type="count"` argument produced a set of histograms that may be somewhat difficult to interpret given the largely unequal N for each smoking habits group and drinking habits group. The unequal N for breakout groups is not a major concern when the `type="percent"` argument is used, however. The histograms based on percent give a sense of the distribution pattern for all four smoking habits groups and all four drinking habits groups. Whether these figures are ever published, these distribution patterns should always be confirmed before the use of any inferential statistical tests.

3.3.5 Density Plot of the Summary Object Variable by Breakout Object Variables

Along with the `lattice::densityplot()` function which is demonstrated in Sample 3, the function `sm::sm.density.compare()` is also used to graphically demonstrate density plots for the breakout object variable. Again, look into the use of density plots, regardless of selected R package.

```
install.packages("sm")
library(sm)                  # Load the sm package.
help(package=sm)             # Show the information page.
sessionInfo()                # Confirm all attached packages.

par(ask=TRUE)
saveline.width <- par(lwd=3) # Generate a heavy line
sm::sm.density.compare(Health.table$SBP,
                       Health.table$Smoke.recode,
  xlab=list("SBP Readings (Limit = 0 to 220)",
  cex=1.15,
  font=2),
  ylab=list("Density", cex=1.15, font=2),
  xlim=c(0,220),
  ylim=c(0,0.25))
title(main="Density Plot of SBP Readings by Smoking  Habits")
colorfill <- c(2:(2+length(levels(Health.table$Smoke. recode))))
legend(locator(1), levels(Health.table$Smoke.recode),
  fill=colorfill)
  # Remember to click on an open location to
  # paste the legend into the figure.
par(saveline.width)
  # Note how the line width is accommodated and then
  # set back to the original value.
```

```
par(ask=TRUE)
saveline.width <- par(lwd=3) # Generate a heavy line
sm::sm.density.compare(Health.table$SBP,
                       Health.table$Drink.recode,
  xlab=list("SBP Readings (Limit = 0 to 220)",
  cex=1.15,
  font=2),
  ylab=list("Density", cex=1.15, font=2),
  xlim=c(0,220),
  ylim=c(0,0.25))
title(main="Density Plot of SBP Readings by Drinking Habits")
colorfill <- c(2:(2+length(levels(Health.table$Drink.recode))))
legend(locator(1), levels(Health.table$Drink.recode),
  fill=colorfill)
  # Remember to click on an open location to
  # paste the legend into the figure.
par(saveline.width)
  # Note how the line width is accommodated and then
  # set back to the original value.
```

In this presentation, the legend is color coded and the legend is generated by clicking the mouse in any desired spot in the graphic. That is to say, the legend is not automatically generated when the graphic is produced. Instead, the graphic is produced and then the user should move over to an open spot and then click the mouse. With a mouse click, the legend will automatically appear. This is a simple demonstration of another way R can be used to embellish a graphical image.

3.3.6 Boxplot of the Summary Object Variable by Breakout Object Variables

```
par(ask=TRUE) # Note breakout group by measured object.
lattice::bwplot(Health.table$Smoke.recode ~
                Health.table$SBP,
  par.settings = simpleTheme(lwd=2),
  par.strip.text=list(cex=1.15, font=2),
  scales=list(cex=1.15),
  main="Boxplot of SBP Readings by Smoking Habits",
  xlab=list("SBP Readings (Limit = 0 to 220)",
  cex=1.15, font=2),
  xlim=c(0,220),
  ylab=list("Smoking Habits", cex=1.15, font=2),
  aspect=0.5,
  layout=c(1,1),
  col="red")
```

```
par(ask=TRUE) # Note breakout group by measured object.
lattice::bwplot(Health.table$Drink.recode ~
                Health.table$SBP,
  par.settings = simpleTheme(lwd=2),
  par.strip.text=list(cex=1.15, font=2),
  scales=list(cex=1.15),
  main="Boxplot of SBP Readings by Drinking Habits",
  xlab=list("SBP Readings (Limit = 0 to 220)",
  cex=1.15, font=2),
  xlim=c(0,220),
  ylab=list("Drinking Habits", cex=1.15, font=2),
  aspect=0.5,
  layout=c(1,1),
  col="red")

par(ask=TRUE)
lattice::bwplot(~ SBP | Smoke.recode * Drink.recode,
  data = Health.table,
  par.settings = simpleTheme(lwd=2),
  par.strip.text=list(cex=1.15, font=2),
  scales=list(cex=1.15),
  main="Two-Way ANOVA:  SBP ~ Smoke + Drink\n
  Smoke = 4 Groups and Drink = 4 Groups",
  xlab=list("SBP Readings (Limit = 0 to 220)\n
  Smoking Habits:
  1 = Never, 2 = No Longer, 3 = 1 to 20, 4 = > 20",
  cex=1.15, font=2),
  xlim=c(0,220),
  ylab=list("Drinking Habits:
  1 = Never, 2 = No Longer, 3 = 1 to 5, 4 = > 5",
  cex=1.15, font=2),
  aspect=0.25,
  layout=c(4,4),
  col="red")
  # The \n character sequence is used to force a
  # new line, which improves presentation.
```

Outliers are identified, if there are any, as the small circles extending beyond the hinges of the boxplots. Outliers are always of interest and demand attention. Is an outlier a possible error in data entry that requires correction? Or is the outlier indeed a correct datum? If so, how is it possible for individual values to deviate so much from norm values? Imagine if the weight of an individual patient were listed as 550 pounds. This weight is certainly possible, but follow-up actions would be prudent to be certain that this weight is correct (as an outlier) and that it is not the result of an error in either measurement or data entry.

When using the `lattice::bwplot()` function, as compared to other uses of the lattice package, the object variable representing breakout groups comes before the object variable representing the measured datum. The two object variables are separated by the tilde (e.g., ~) character.

3.4 Descriptive Analysis of the Data

The many figures prepared up to this place in the lesson should provide a fairly good vision of the data that are of importance to this study, individually and in tandem with each other. A full set of descriptive statistics, both summary descriptive statistics and breakout descriptive statistics, will help round out a more complete understanding of the data and support the eventual two-way ANOVA of SBP by smoke and by drink.

3.4.1 Summary Descriptive Statistics

R functions typically used to support descriptive statistics at the summary level include: `mean()`, `median()`, `summary()`, `psych::describe()`, and the exceptionally useful `epicalc::summ()` function.

```
mean(Health.table$SBP, na.rm=TRUE)
median(Health.table$SBP, na.rm=TRUE)
summary(Health.table$SBP)

par(ask=TRUE)
epicalc::summ(Health.table$SBP, # Use the epicalc package.
  by=NULL,
  graph=TRUE,
  box=TRUE,        # Generate a boxplot.
  pch=18,
  ylab="auto",
  main="Sorted Dotplot and Boxplot of Health.table$SBP",
  cex.X.axis=1.15,
  cex.Y.axis=1.15,
  font.lab=2,
  dot.col="auto")
```

The `epicalc::summ()` by=NULL argument purposely sets output so that breakout statistics are not provided. Instead, for this sample, descriptive statistics and an accompanying graphic for all values in `Health.table$SBP` are provided.

```
install.packages("psych")
library(psych)              # Load the psych package.
```

```
help(package=psych)      # Show the information page.
sessionInfo()            # Confirm all attached packages.
```

```
psych::describe(Health.table$SBP)
```

The psych::describe() function is applied against the entire dataset. The desired output can then be copied and pasted into a formal report. Merely delete the output applied against numbers that are instead factor-type object variables, which are marked with a * symbol.

```
psych::describe(Health.table)
```

A similar type of output is obtained with the epicalc::summ() function, when applied against the full set of data, organized in the data frame. Attention should always be given to quality assurance checks, so use this function to look at the min (minimum) and max (maximum) output, to see if there are data that are simply out-of-range (e.g., a 0, 6, etc. showing as min or max output for data from a 1 to 5 Likert-type scale).

```
epicalc::summ(Health.table)
```

3.4.2 Breakout Descriptive Statistics

Descriptive statistics at the summary level are needed, but with an emphasis on two-way ANOVA breakout statistics for the different smoke groups and drink groups are also needed to gain a more complete understanding of the data. From among the many ways to obtain breakout statistics, the tapply() function, the psych::describe.by() function, the epicalc::summ() function, the doBy::summaryBy(), and the prettyR::brkdn() function may represent the best selections to gain a better sense of differences for the four breakout groups for the factor object variable Health.table$Smoke.recode (never smoked, no longer smokes, 1–20 cigarettes a day, and greater than 20 cigarettes a day) and the four breakout groups for the factor object variable Health.table$Drink.recode (never drank, no longer drinks, 1–5 drinks a week, and greater than 5 drinks a week).

Compare the output gained by use of these many functions, to see which ones meet reporting requirements and ease of transfer by simple copy/paste.

```
# tapply() Function
tapply(SBP, Smoke.recode, summary, na.rm=TRUE,
  data=Health.table)
tapply(SBP, Drink.recode, summary, na.rm=TRUE,
  data=Health.table)
```

```
#psych::describe.by() Function
psych::describe.by(Health.table$SBP,
                    Health.table$Smoke.recode)
psych::describe.by(Health.table$SBP,
                    Health.table$Drink.recode)

# epicalc::summ() Function
par(ask=TRUE)
epicalc::summ(Health.table$SBP,
   by=Health.table$Smoke.recode,   # Note the use of by.
   graph=TRUE, # Use graph=TRUE,
   pch=20,      # if desired.
   ylab="auto",
   main="Sorted Dotplot of Health.table$SBP
   by Health.table$Smoke.recode",
   cex.X.axis=1.125, # Adjust presentation to best effect.
   cex.Y.axis=1.125, # Adjust presentation to best effect.
   font.lab=2,
   dot.col="auto")

# epicalc::summ() Function
par(ask=TRUE)
epicalc::summ(Health.table$SBP,
   by=Health.table$Drink.recode,
   graph=TRUE, # Use graph=TRUE,   # Note the use of by.
   pch=20,      # if desired.
   ylab="auto",
   main="Sorted Dotplot of Health.table$SBP
   by Health.table$Drink.recode",
   cex.X.axis=1.125, # Adjust presentation to best effect.
   cex.Y.axis=1.125, # Adjust presentation to best effect.
   font.lab=2,
   dot.col="auto")

# doBy::summaryBy() Function
install.packages("doBy")
library(doBy)              # Load the doBy package.
help(package=doBy)         # Show the information page.
sessionInfo()              # Confirm all attached packages.
```

From among the many possible functions available for breakout descriptive statistics, the doBy::summaryBy() function is also useful, providing information in a fairly condensed format, similar to what might be expected from a crosstabs analysis.

```
doBy::summaryBy(SBP ~ Smoke.recode +
                Drink.recode,
                data=Health.table,
```

```
    FUN=c(mean,sd),
    na.rm=TRUE,
    keep.names=TRUE,
    order=TRUE)

doBy::summaryBy(SBP ~ Drink.recode +
                      Smoke.recode,
                      data=Health.table,
    FUN=c(mean,sd),
    na.rm=TRUE,
    keep.names=TRUE,
    order=TRUE)
```

It will be difficult to accommodate missing values for length. The enumerated function below takes care of this problem.

```
descriptivefun <- function(x, ...){
    c(m=mean(x, ...), sd=sd(x, ...), l=length(x))
    } # This is a user-created (e.g., enumerated) function.

doBy::summaryBy(SBP ~ Smoke.recode +
                      Drink.recode,
                      data=Health.table,
    FUN=descriptivefun,
    na.rm=TRUE,
    keep.names=TRUE,
    order=TRUE)
    # Use the enumerated function.

doBy::summaryBy(SBP ~ Drink.recode +
                      Smoke.recode,
    data=Health.table,
    FUN=descriptivefun, # Use the enumerated function.
    na.rm=TRUE,
    keep.names=TRUE,
    order=TRUE)
    # Use the enumerated function.

# prettyR::brkdn() Function

install.packages("prettyR")
library(prettyR)          # Load the prettyR package.
help(package=prettyR)     # Show the information page.
sessionInfo()             # Confirm all attached packages.

prettyR::brkdn(SBP ~ Smoke.recode,
    data=Health.table,
    maxlevels=4,
```

```
    num.desc=c("mean", "sd", "valid.n"),
    width=25,
    round.n=2)

prettyR::brkdn(SBP ~ Drink.recode,
    data=Health.table,
    maxlevels=4,
    num.desc=c("mean", "sd", "valid.n"),
    width=25,
    round.n=2)

prettyR::brkdn(SBP ~ (Smoke.recode +
                      Drink.recode),
    data=Health.table,
    maxlevels=4,
    num.desc=c("mean","sd","valid.n"),
    width=25,
    round.n=2)
    # Observe how the summary statistic
    # is provided, and then breakouts.

prettyR::brkdn(SBP ~ (Drink.recode +
                      Smoke.recode),
    data=Health.table,
    maxlevels=4,
    num.desc=c("mean","sd","valid.n"),
    width=25,
    round.n=2)
    # Observe how the summary statistic
    # is provided, and then breakouts.
```

With these descriptive statistics now generated at the breakout level, look at the way they can be incorporated into one fairly descriptive graphical image, for SBP readings for all smoke breakout groups and also for all drink breakout groups. A color-coded legend and a descriptive table, presented as an additional legend, are also included in this graphical image (see Fig. 3.1).

```
# plotrix::brkdn.plot() Function
install.packages("plotrix")  .
library(plotrix)        # Load the plotrix package.
help(package=plotrix)   # Show the information page.
sessionInfo()           # Confirm all attached packages.
```

In the next figure, look at the way par() settings are toggled specific to the figure. A setting such as making the font show in "bold" is set before the figure and then the setting is set back to the default at the end of the figure. This is a good technique to use when working with graphical images, when R arguments in the package do not always support desired output. With some level of trial-and-error, R can usually be set to produce desired output.

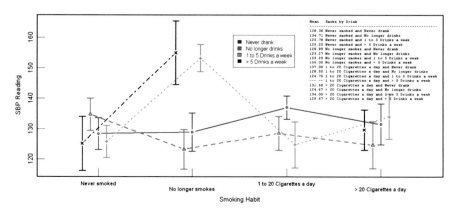

Fig. 3.1 Breakdown plot of SBP readings by factors Smoke.recode and Drink .recode (Mean): plotrix::brkdn.plot ()

```
savelwd         <- par(lwd=3)        # Heavy line
savefont        <- par(font=2)       # Bold
savecex.lab     <- par(cex.lab=1.25) # Label
savecex.axis    <- par(cex.axis=1.25)# Axis
savetck         <- par(tck=0.05)     # Tick Marks
par(ask=TRUE)
plotrix::brkdn.plot("SBP", "Drink.recode",
                             "Smoke.recode",
  Health.table,
  mct="mean",
  dispbar=TRUE,
  main="Breakdown Plot of SBP Readings by Factors
  Smoke.recode and Drink.recode (Mean):
  plotrix::brkdn.plot()",
  xlab="Smoking Habit",
  ylab="SBP Reading",
  type="b",
  pch=1:4,
  lty=1:4,
  col=c("blue","red","green", "black"),
  staxx=TRUE)
par(savelwd)         # Go back to default settings.
par(savefont)        # Go back to default settings.
par(savecex.lab)     # Go back to default settings.
par(savecex.axis)    # Go back to default settings.
par(savetck)         # Go back to default settings.
colorfill <- c("blue","red","green","black")
# Place the mouse over the desired location for
# the legend and then left click to drop the
# legend into the figure.
legend(locator(1),
```

```
    levels(Health.table$Drink.recode),
    fill=colorfill)
    # Give attention to the color ordering.
    # Never assume that the ordering reflects
    # the values but instead confirm that the
    # colors reflect true values.
savefamily <- par(family = "mono")        # Courier-type font.
legend("topright",
    legend = c(
    "Mean    Smoke by Drink                            ",
    "---------------------------------------------------",
    "128.36 Never smoked and Never drank               ",
    "134.71 Never smoked and No longer drinks          ",
    "125.78 Never smoked and 1 to 5 Drinks a week      ",
    "125.20 Never smoked and > 5 Drinks a week         ",
    "128.89 No longer smokes and Never drank           ",
    "123.27 No longer smokes and No longer drinks      ",
    "153.20 No longer smokes and 1 to 5 Drinks a week ",
    "155.00 No longer smokes and > 5 Drinks a week     ",
    "137.00 1 to 20 Cigarettes a day and Never drank   ",
    "128.50 1 to 20 Cigarettes a day and No longer drinks    ",
    "124.75 1 to 20 Cigarettes a day and 1 to 5 Drinks a week",
    "---.-- 1 to 20 Cigarettes a day and > 5 Drinks a week   ",
    "131.50 > 20 Cigarettes a day and Never drank      ",
    "124.67 > 20 Cigarettes a day and No longer drinks",
    "134.00 > 20 Cigarettes a day and 1 to 5 Drinks a week   ",
    "129.67 > 20 Cigarettes a day and > 5 Drinks a week      "),
    ncol=1,
    locator(1),
    xjust=1,
    text.col="black",
    cex=0.70,
    inset=0.01,
    bty="n")
par(savefamily)
```

The legend for this figure is rather large and it depends on column ordering for best presentation. It was set to a Courier-type mono-spaced font using the family = "mono" par setting to have all Mean values and character values line up properly.

A color-coded legend is of limited value to anyone who has trouble distinguishing one color from another. Plus, the details of the text-based descriptive legend provide even greater detail than a color-coded legend can provide. Of course, the figure should not be so busy that it is difficult to read and understand.

Now that the collapsed and breakout statistics are completed, the data are ready for further analyses, such as two-way ANOVA for this sample.

3.5   Use R for Two-Way Analysis of Variance (ANOVA)

The many prior actions for this lesson may seem redundant at first, but they are a necessary part of the research process. It is not uncommon to devote up to 70% (or more) of all time-on-task for statistical analyses to data preparation and quality assurance. Experienced researchers know the importance of attention to data preparation, data scrubbing, descriptive statistics as a quality assurance tool, and graphical presentations as an advance organizer to inferential analyses. Only an inexperienced researcher would rush into an inferential analysis without going through these preparatory actions.

The task now is to use the many functions supported by R to determine if there are statistically significant differences between (1) the four smoking habits breakout groups, (2) the four drinking habits breakout groups, and (3) interaction between smoking habits and drinking habits. Two-way ANOVA will be used to support this task.

R supports many possible ways to perform a two-way ANOVA. From among many possible approaches, a few methods that support two-way ANOVA are detailed below.

3.5.1   Two-Way ANOVA: aov () Function

Use the formula for a two-way factorial design ANOVA, which is typically represented as:

```
fit1 <- aov(y ~ A + B + A:B, data=dataframe)
summary(fit1)

fit2 <- aov(y ~ A*B, data=dataframe)
summary(fit2)

y   = Measured datum, (e.g., weight, exam score, etc.)
A   = Factor Variable A (e.g., Gender, Race-Ethnicity, etc.)
B   = Factor Variable B (e.g., Soil Type, Breed Type, etc.)
A:B = Interaction of Factor Variable A and Factor Variable B
```

Both ANOVA formulas yield the same result, but the following formula (e.g., algorithm) is likely more illustrative:

```
fit1 <- aov(y ~ A + B + A:B, data=dataframe)

Sample2.fit1 <- aov(SBP ~ Smoke.recode +
                         Drink.recode +
                         Smoke.recode:Drink.recode,
                data=Health.table)
```

```
summary(Sample2.fit1)

#                         Df Sum Sq Mean Sq F value Pr(>F)
# Smoke.recode            3    412  137.30  0.3992 0.75385
# Drink.recode            3    449  149.80  0.4356 0.72802
# Smoke.recode:
#    Drink.recode         8   6180  772.48  2.2461 0.02983 *
# Residuals             102  35080  343.92
# ---
# 3 observations deleted due to missingness
# >

Sample2.fit2 <- aov(SBP ~ Smoke.recode*Drink.recode,
              data=Health.table)
summary(Sample2.fit2)

#                         Df Sum Sq Mean Sq F value Pr(>F)
# Smoke.recode            3    412  137.30  0.3992 0.75385
# Drink.recode            3    449  149.80  0.4356 0.72802
# Smoke.recode:
#    Drink.recode         8   6180  772.48  2.2461 0.02983 *
# Residuals             102  35080  343.92
# ---
# 3 observations deleted due to missingness
# >
```

To gain a sense of the descriptive statistics (summary and breakout), use the model.tables() function for another view of Grand Mean, Mean, and *N* for each cell in the factorial table.

```
print(model.tables(Sample2.fit1,"means"), digits=3)
print(model.tables(Sample2.fit2,"means"), digits=3)
```

As a throwaway diagnostic, use the plot.design() function to see general trends for SBP readings by each breakout group.

```
par(ask=TRUE)
plot.design(SBP ~ Smoke.recode + Drink.recode,
  data=Health.table)
```

Then, use the interaction.plot() function for more details, but now focusing on the object variables of greatest interest.

```
par(ask=TRUE)
interaction.plot(Health.table$Smoke.recode,
                 Health.table$Drink.recode,
                 Health.table$SBP,
  main="Interaction Plot: Smoking Habits, Drinking Habits,
  and SBP",
  font.lab=2, col=2:9, lty="solid", lwd=6)
```

The above `aov()` function more than meets the basic need for a two-way ANOVA. However, there are many other R functions that can go well beyond this basic information. A few of these other R functions will be demonstrated below. These additional R functions provide a rich understanding of the data and they provide additional insight into interaction(s). These insights should not be overlooked and are always of interest to experienced researchers. Indeed, these insights are perhaps the main purpose for engaging in an inferential test. Otherwise, descriptive statistics alone would likely suffice if insight were not desired.

3.5.2   s20x Package

The s20x package provides many other functions that assist with two-way ANOVA and a more complete understanding of the data and subsequent outcomes related to analysis of variance.

```
install.packages("s20x")
library(s20x)             # Load the s20x package.
help(package=s20x)        # how the information page.
sessionInfo()             # Confirm all attached packages.

saveaxis <- par(cex.axis=0.65)
par(ask=TRUE)
s20x::boxqq(SBP ~ Smoke.recode, data=Health.table)
par(saveaxis)
   # Labels on the X axis were made smaller so that
   # all text shows in the figure.

saveaxis <- par(cex.axis=0.65)
par(ask=TRUE)
s20x::boxqq(SBP ~ Drink.recode, data=Health.table)
par(saveaxis)
   # Labels on the X axis were made smaller so that
   # all text shows in the figure.

s20x::crosstabs(~ Smoke.recode + Drink.recode, Health.table)
```

Going back to Sample2.fit1 and Sample2.fit2, the `s20x::summary2way()` function can be used to generate a two-way ANOVA table. Give attention to the additional descriptive information provided with use of the `s20x::summary2way()` function, as opposed to use of the `aov()` function alone.

```
s20x::summary2way(Sample2.fit1,
   page = "table",
   digit = 5,
   conf.level = 0.95,
   print.out = TRUE)
```

```
s20x::summary2way(Sample2.fit1,
  page = "means",
  digit = 5,
  conf.level = 0.95,
  print.out = TRUE)

s20x::summary2way(Sample2.fit2,
  page = "table",
  digit = 5,
  conf.level = 0.95,
  print.out = TRUE)

s20x::summary2way(Sample2.fit2,
  page = "means",
  digit = 5,
  conf.level = 0.95,
  print.out = TRUE)
```

The page="means" argument provides detailed summary statistics for overall, by the different factor object variables, and by combinations of the different factor object variables. This argument is rich and it should be used whenever this type of analysis is attempted.

3.5.3   Outcome to Sample 2

A variety of R functions have been used to import data, organize data, label data, provide a sense of summary and breakout descriptive statistics, and provide a graphical view of the data at the summary level and breakout level. Different R functions were then used to conduct the two-way ANOVA inferential test of systolic blood pressure by smoking habits and by drinking habits.

Given its importance, consider once again the null hypothesis for this sample lesson: There is no difference between smoking habits, drinking habits, and interaction between smoking habits and drinking habits regarding systolic blood pressure readings from among patients in a convenience sample (p <= 0.05). As prepared, this null hypothesis addresses three separate, but related questions: (1) In regard to measured systolic blood pressure (SBP), are the population means for smoking habits (the first factor object variable in question; never smoked, no longer smokes, 1–20 cigarettes a day, and greater than 20 cigarettes a day) equal? (2) In regard to measured systolic blood Pressure (SBP), are the population means for drinking habits (the second factor object variable in question; never drank, no longer drinks, 1–5 drinks a week, and greater than 5 drinks a week) equal? (3) In regard to measured systolic blood pressure (SBP), is there any observable interaction between smoking habits and drinking habits?

Consider again the two-way ANOVA of systolic blood pressure (SBP) by smoking habits and by drinking habits, as generated by the fit-type analyses:

```
#                       Df Sum Sq Mean Sq F value Pr(>F)
# Smoke.recode           3    412  137.30  0.3992 0.75385
# Drink.recode           3    449  149.80  0.4356 0.72802
# Smoke.recode:
#   Drink.recode         8   6180  772.48  2.2461 0.02983 *
# Residuals            102  35080  343.92
# ---
# 3 observations deleted due to missingness
# >
```

By reviewing this output and using p <= 0.05 as the criterion *p*-value for the null hypothesis, observe how:

3.5.3.1 Smoking Habits

There is no difference in Mean SBP readings by the four breakout smoking habits (calculated p <= 0.75 which exceeds the criterion p <= 0.05). The descriptive statistics that support this observation are from the prettyR ::brkdn() function, using copy and paste with minor editing:

```
# Breakdown of SBP by Smoke.recode
#                        Level   mean     sd  valid.n
#                 Never smoked  129.6  18.02       39
#            No longer smokes  132.7  23.01       31
# 1 to 20 Cigarettes a day     131.0  16.32       22
#     > 20 Cigarettes a day    131.3  19.07       27
```

3.5.3.2 Drinking Habits

There is no difference in Mean SBP readings by the four breakout drinking habits (calculated p <= 0.73 which exceeds the criterion p <= 0.05). The descriptive statistics that support this observation are from the prettyR:: brkdn() function, using copy and paste with minor editing:

```
# Breakdown of SBP by Drink.recode
#               Level     mean     sd  valid.n
#         Never drank     131.4  16.59       38
#    No longer drinks     129.1  19.18       32
# 1 to 5 Drinks a week    132.3  20.69       33
#     > 5 Drinks a week   134.9  21.47       15
```

3.5.3.3   Interaction Between Smoking Habits and Drinking Habits

In view of SBP readings, the interaction effect between smoking habits and drinking habits was significant (calculated `p <= 0.03` (`0.02983` to be more precise)), which is less than the criterion `p <= 0.05`). As a discussion point that strays from the null hypothesis, is the interaction effect between smoking habits and drinking habits significant at `p <= 0.01`?

Consider the descriptive statistics gained from a copy and paste action (after only minor editing) using the `s20x::summary2way()` function with further use of the `page="means"` argument:

```
#                                    1 to 5
#                  Never  No longer Drinks a > 5  Drinks
#                  drank  drinks    week     a week      Total
# Never smoked     128.36 134.71    125.78   125.20      128.51
# No longer smokes 128.89 123.27    153.20   155.00      140.09
# 1 to 20 Cigarettes
#    a day         137.00 128.50    124.75   NaN         131.02
# > 20 Cigarettes
#    a day         131.50 124.67    134.00   129.67      129.96
# Total            131.44 127.79    134.43   135.93      132.40
```

A careful examination of this table provides possible evidence of how to react to this finding of significant interaction. Specifically:

There were no subjects who met the category of `"1 to 20 cigarettes a day"` and `"Greater than 5 drinks a week"`. Empty cells generally make it difficult to provide firm conclusions and again, it is reminded that this dataset was from an inexperienced student. Seasoned researchers may have had better outcomes on the issue of missing data and an empty cell, possibly by obtaining a sample that was more representative of the population.

Merely using casual observation of cell means, note how means for (1) `"No longer smokes"` and `"1 to 5 drinks a week"` (Mean = 153.20) and (2) `"No longer smokes"` and `"Greater than 5 drinks a week"` (Mean = 155.00) seem to greatly exceed the Grand Mean (Mean = 131.05).

To supplement this casual observation of cell means, look also at the figures that were previously generated and repeated below, using the `plotrix::brkdn.plot()` function.

```
install.packages("plotrix")
library(plotrix)        # Load the plotrix package.
help(package=plotrix)   # Show the information page.
sessionInfo()           # Confirm all attached packages.

savelwd       <- par(lwd=3)        # Heavy line
savefont      <- par(font=2)       # Bold
savecex.lab   <- par(cex.lab=1.25) # Label
savecex.axis  <- par(cex.axis=1.25)# Axis
```

```
savetck        <- par(tck=0.05)       # Tick Marks
par(ask=TRUE)
plotrix::brkdn.plot("SBP", "Drink.recode",
                               "Smoke.recode",
    Health.table,
    mct="mean",
    dispbar=TRUE,
    main="Breakdown Plot of SBP Readings by Factors
    Smoke.recode and Drink.recode (Mean):
    plotrix::brkdn.plot()",
    xlab="Smoking Habit",
    ylab="SBP Reading",
    type="b",
    pch=1:4,
    lty=1:4,
    col=c("blue","red","green", "black"),
    staxx=TRUE)
par(savelwd)       # Go back to default settings.
par(savefont)      # Go back to default settings.
par(savecex.lab)   # Go back to default settings.
par(savecex.axis)  # Go back to default settings.
par(savetck)       # Go back to default settings.
colorfill <- c("blue","red","green","black")
legend(locator(1),
    levels(Health.table$Drink.recode),
    fill=colorfill)
    # Give attention to the color ordering.
    # Never assume that the ordering reflects
    # the values but instead confirm that the
    # colors reflect true values.
savefamily <- par(family = "mono")    # Courier-type font.
legend("topright",
    legend = c(
    "Mean    Smoke by Drink                          ",
    "------------------------------------------------",
    "128.36 Never smoked and Never drank             ",
    "134.71 Never smoked and No longer drinks        ",
    "125.78 Never smoked and 1 to 5 Drinks a week    ",
    "125.20 Never smoked and > 5 Drinks a week       ",
    "128.89 No longer smokes and Never drank         ",
    "123.27 No longer smokes and No longer drinks    ",
    "153.20 No longer smokes and 1 to 5 Drinks a week",
    "155.00 No longer smokes and > 5 Drinks a week   ",
    "137.00 1 to 20 Cigarettes a day and Never drank ",
    "128.50 1 to 20 Cigarettes a day and No longer drinks    ",
    "124.75 1 to 20 Cigarettes a day and 1 to 5 Drinks a week",
    "---.-- 1 to 20 Cigarettes a day and > 5 Drinks a week   ",
    "131.50 > 20 Cigarettes a day and Never drank    ",
    "124.67 > 20 Cigarettes a day and No longer drinks",
```

```
   "134.00 > 20 Cigarettes a day and 1 to 5 Drinks a week      ",
   "129.67 > 20 Cigarettes a day and > 5 Drinks a week         "),
     ncol=1,
     locator(1),
     xjust=1,
     text.col="black",
     cex=0.70,
     inset=0.01,
     bty="n")
par(savefamily)
```

The last figure, based on the `plotrix::brkdn.plot()` function applied against the object variables SBP, `Drink.recode`, and `Smoke.recode` is perhaps the most telling. The magnitude of difference of SBP readings for (1) "No longer smokes" and "1 to 5 drinks a week" (Mean SBP = 153.20) and (2) "No longer smokes" and "Greater than 5 drinks a week" (Mean SBP = 155.00) is clearly evident in this figure and the magnitude of difference is reinforced by careful review of Mean SBP readings in the legend.

Accordingly, with a fair degree of certainty, the sample provided evidence that, for this dataset, smoking had no direct measurable impact on SBP readings, drinking had no direct measurable impact on SBP readings, but there was an observed interaction between smoking and drinking on SBP readings. Specifically, SBP values were highest for those individuals who no longer smoke but drink alcohol, whether "1 to 5 drinks a week" (Mean SBP = 153.20) or "Greater than 5 drinks a week" (Mean SBP = 155.00). As such, it can be said that the effect of drink on SBP was greatest among those who no longer smoke.

These conclusions are stated with the constant caution that tight controls and replication are both essential parts of the overall research process. Consistent outcomes across various scenarios, various subjects, finite control of operational definitions for smoking habits and drinking habits, would be needed before any definitive conclusions could be offered. This sample is merely one attempt in this overall assessment of factors impacting public health.

This sample was a useful example of interaction between chemicals, which is certainly a public health issue. Consider the detailed history often asked by both medical staff and then again by pharmacists before a new prescription is placed. Concern about interaction is a major motivation for what may at first seem to be overly obtrusive questions about diet, prescribed medicines, over-the-counter medicines, vitamins, herbal supplements, etc. Combinations of two or more chemicals can cause a profound interaction and two-way ANOVA is one possible selection to examine these effects.

```
############################################################
Prepare to Exit, Save, and Later Retrieve This R Session #
############################################################
getwd()            # Identify the current working directory.
ls.str()           # List all objects, with finite detail.
```

```
save.image("Two-Way_ANOVA_Smoke-and-Drink.rdata")
list.files()        # List files at the PC directory.
q()                 # Quit this session.
                    # Prepare for Save workspace image? query.
###################### END #############################
```

Use the R graphical user interface (GUI) to load the saved .rdata file: File and then load workspace. Otherwise, use the load() function, keying the full pathname, to load the .rdata file and retrieve the session. Recall, however, that it may be just as useful to simply use the .R script file and recreate the analyses and graphics, provided the data files remain available.

Chapter 4
Two-Way Analysis of Variance (ANOVA) Sample 3: Comparison of Larvae Counts by AgChem Formulation and by AgChem Application Time-of-Day

Abstract In this chapter, R is used to support detailed analyses relevant to the biological sciences, where main effects and possible interactions are examined for how a specific insect reacts to different chemical formulations and environmental factors. New R-based tools (packages and associated functions and graphics) are introduced in this chapter, resulting in graphical output that approaches publishable quality. An additional feature in this chapter is the presentation of R syntax that treats data from a nonparametric perspective, which should be considered if there are concerns about data distribution patterns or other assumptions inherent to selected tests. This chapter concludes with optional suggestions on how to manage the R work session for lengthy and complex analyses.

Integrated Pest Management (IPM) represents a balanced attempt to counter the yield-reducing impact of agricultural pests (e.g., weeds, insects, disease-causing pathogens, etc.). Producers who practice IPM use Agricultural Chemicals (e.g., AgChems), but IPM attempts to limit the use of AgChems by applying knowledge of pest life cycles, weather conditions, AgChem properties, etc. to optimum effect. IPM field scouts play a key role in this balanced approach to crop production. IPM scouts are the individuals who actually go into the field and closely monitor, record, and transmit field conditions such as: ambient temperature and humidity at field ground level, sunlight at field ground level, number and type of insects at random locations throughout a field, internodal length of new growth, etc. This base information is then used to decide if pesticide applications are warranted, given the desire for optimum yield and not only maximum yield. Within the context of IPM, the decision-making process for AgChem application takes into account the cost of application if AgChems are used versus the cost of reduced yield if AgChems are not used.

In an attempt to minimize any potential for bias for this sample lesson, the principal analyst knows only that the data represent: (1) Formula (e.g., three formulations of an AgChem), (2) Time-of-Day (e.g., early morning (AM) application of these three AgChems versus late afternoon (PM) application of these three AgChems), and (3) Larvae (e.g., number of larvae for a specific insect per square meter at random locations one week after AgChem application.

The scenario for how the data were gained for this sample represents a highly controlled set of actions: (1) A large field of uniform soil type and topology was divided into six ten acre sections, with large mowed grass alleyways between each section, serving as a buffer area between the different ten acre field sections. (2) An AgChem was prepared and applied on separate sections both in the early morning (AM) and again about ten hours later (PM) using three formulations by two time periods (AgChem Formula 1 AM and AgChem Formula 1 PM, AgChem Formula 2 AM, and AgChem Formula 2 PM, and AgChem Formula 3 AM, and AgChem Formula 3 PM). For context, the active ingredients in the three AgChems formulations are the same and the active ingredients are applied at the same rate (e.g., dose). However, the formulations are different. Some AgChems are prepared with buffering agents, detergents, dispersants, emulsifiers, wetting agents, etc. Some AgChems are prepared as liquids (often of various viscosities), some are prepared as wettable powders, some are prepared as granules. The principal analyst only knows the three formulations as: AgChem Formula 1, AgChem Formula 2, and AgChem Formula 3. Equally, there may be difference in chemical reaction due to environmental conditions (e.g., sunlight, humidity, etc.) possibly related to time-of-day application. For this sample, the two application times are early morning (AM) and late afternoon (PM). (3) A week later, a team of IPM scouts from an 1890 Morrill Act Land Grant university went into the field and following strict protocols counted the number of larvae (without regard to larvae size or weight) for a specific insect at 50 random locations for each of the six ten acre sections, resulting in a dataset of 300 (50 counts per ten acre field section × 6 ten acre field sections) cases.

This study examines if there are differences in the number of larvae for an unidentified (to the principal analyst) insect by AgChem formula (three breakout groups), by AgChem time-of-day application (two breakout groups), and any possible interaction between AgChem formula and AgChem time-of-day application.

For this sample lesson, AgChem formula and AgChem time-of-day application will be viewed as nominal data. There is no ordering to either the three AgChem formulas or the two AgChem time-of-day application times.

With these declarations established, two-way analysis of variance (ANOVA) was judged to be the appropriate test for this factorial-type analysis of summative differences in larvae counts (larvae counts are considered an acceptable proxy for weight, given how all larvae were collected within a few hours of each other) by AgChem formula (three breakout groups) and AgChem time-of-day application (two breakout groups). The header row, the first three lines of data, and the last three lines of data for the .csv (comma-separated values) file are shown below, with commas separating values in each row:

```
ID,Formula,TimeOfDay,Larvae
68,1,1,130
76,1,1,130
```

```
79,1,1,131
9920,3,2,137
9946,3,2,137
9970,3,2,137
```

Sample 3 Ho (Null Hypothesis): There is no difference between AgChem formula, AgChem application time-of-day, and interaction between AgChem formula and AgChem application time-of-day regarding larvae counts for an unidentified insect at randomly selected areas in a field with otherwise uniform features (p $\leq$ 0.05).

4.1 Data Import of a .csv Spreadsheet-Type Data File into R

The data were originally recorded on tablet-type appliances issued to the IPM scouts. The lead IPM scout reviewed the data and when the data were considered acceptable they were transferred to a PC and saved as an open source Gnumeric-based spreadsheet, in .gnumeric file format. This action adheres to the desire to use low cost or no cost software that is still of the highest quality. To promote ease of future data distribution, the data were then saved in .csv file format. Because the data are in .csv file format, it should be a fairly easy action to import the data into R. The data for this lesson, in comma-separated values format as an ASCII .csv file, have been placed at the F:\R_Lessons\Inferential_Statistics_Parametric directory on a standalone personal computer.

All analyses begin from this starting point, working with a previously prepared .csv format ASCII-text file. The emphasis will be on: (1) overall analysis of Larvae counts, (2) breakout analyses of Larvae counts: Larvae counts by AgChem formula and Larvae counts by AgChem application time-of-day, (3) graphical representation of overall and breakout findings, (4) two-way ANOVA of the data, and (5) summative interpretation of outcomes.

Note below how the read.table() function is used to read in the comma-separated values .csv format ASCII file. This sample continues the demonstration of how R can accommodate multiple file formats and file extensions.

```
################################################################
# Housekeeping              Use for All Analyses
################################################################
date()             # Current system time and date.
R.version.string   # R version and version release date.
options(digits=5)  # Confirm default digits.
options(scipen=999)# Suppress scientific notation.
options(width=60)  # Confirm output width.
ls()               # List all objects in the working
                   # directory.
rm(list = ls())    # CAUTION:  Remove all files in the
                   #           working directory.  If this
                   #           action is not desired, use rm()
```

```
                    #              one-by-one to remove the objects
                    #              that are not needed.
ls.str()            # List all objects, with finite detail.
getwd()             # Identify the current working directory.
setwd("F:/R_Lessons/Inferential_Statistics_Parametric")
                    # Set to a new working directory.
                    # Note the single forward slash and double
                    # quotes.
                    # This new directory should be the directory
                    # where the data file is located, otherwise
                    # the data file will not be found.
getwd()             # Confirm the working directory.
list.files()        # List files at the PC directory.
.libPaths()         # Library pathname.
.Library            # Library pathname.
sessionInfo()       # R version, locale, and packages.
search()            # Attached packages and objects.
searchpaths()       # Attached packages and objects.
################################################################
```

4.1.1   Data Import or Data Entry

```
AgChem.table <- read.table (file =
  "AgChem_Formula_Time-of-Day.csv",
  header = TRUE,
  sep = ",")             # Import the .csv file.
getwd()                  # Identify the working directory.
ls()                     # List objects.
attach(AgChem.table)     # Attach the data, for later use.
names(AgChem.table)      # Identify names.
head(AgChem.table)       # Show the head.
tail(AgChem.table)       # Show the tail.
AgChem.table             # Show the entire data frame.
```

As part of any new set of analyses with R, be sure that there are no prior R-based datasets called AgChem.table available. Note how it was only necessary to key the filename for the .csv file and not the full pathname since the R working directory is currently set to the directory and subdirectory where this .csv file is located (see Sect. 4.1 at the beginning of this lesson).

4.2   Organize the Data and Display the Code Book

This dataset is fairly simple, consisting of only four object variables: ID, formula, timeofday, and larvae. Immediately after the data have been imported into R, it is usually necessary to check the data for format and then make any changes that may be needed, to organize the data. As a typical example, consider how numeric codes have been used to identify factors (e.g., groups) in this dataset, specifically the numeric codes for object variables formula and TimeOfDay.

Comment and Warning: R is case sensitive regarding names for object variables. The O in TimeOfDay is in CAPS and in R, TimeOfDay (which is the correct name for this object variable) is not the same as TimeofDay (which is not the correct name for this object variable).

```
class(AgChem.table)
class(AgChem.table$ID)          # DataFrame$ObjectName notation.
class(AgChem.table$Formula)     # DataFrame$ObjectName notation.
class(AgChem.table$TimeOfDay)   # DataFrame$ObjectName notation.
class(AgChem.table$Larvae)      # DataFrame$ObjectName notation.
```

With the class() function applied against each object, consult the code book and coerce each object, as needed, into its correct class. Typically, integers that serve as numeric codes (e.g., 1 represents Alfalfa and 2 represents Red Clover) are coerced into factor format.

However, before the object variables are coerced into correct type, it is a good idea to check the ID object variable for duplicates. A random process was used to assign an ID number to each case and sometimes duplicate numbers are used. This should not happen, but experienced researchers anticipate these problems and then use simple tools, prior to analyses, to make corrections. The duplicated() function is used to check for any duplicates.

```
duplicated(AgChem.table$ID) # Look for FALSE in each case.
```

It may also be a good idea to prepare a quick plot of each object variable, as a throwaway graphic that is only intended to immediately look for anything that may seem illogical or out-of-range.

```
par(ask=TRUE)
plot(AgChem.table$ID)           # Quick diagnostic.

par(ask=TRUE)
plot(AgChem.table$Formula)      # Quick diagnostic.

par(ask=TRUE)
plot(AgChem.table$TimeOfDay)    # Quick diagnostic.

par(ask=TRUE)
plot(AgChem.table$Larvae)       # Quick diagnostic.

# Code Book ###################################
#############################################
# Integrated Pest Management (IPM) Code Book
#
# Variable Labels
#   ID            Identification Number
#   Formula       AgChem (Agricultural Chemical) Formula
#   TimeOfDay     Time-of-Day of AgChem Application
```

```
#    Larvae         Number of Larvae
#
# Variable Values
#    ID             Nominal    LOW to HIGH
#    Formula        Nominal    1 AgChem Formula 1
#                              2 AgChem Formula 2
#                              3 AgChem Formula 3
#    TimeOfDay      Nominal    1 AM
#                              2 PM
#    Larvae         Interval   LOW to HIGH
####################################################
```

To follow a conservative approach, the factor object variables in this dataset are declared as nominal values. It is assumed that there is no meaningful ordering to either Formula or TimeOfDay.

```
install.packages("epicalc")
library(epicalc)           # Load the epicalc package.
help(package=epicalc)      # Show the information page.
sessionInfo()              # Confirm all attached packages.
```

Documentation is an essential part of the research process. Given this aim and in an effort to also promote readability, all object variables should be labeled. Although there are different ways R can be used to meet this goal, the epicalc::label.var() function can easily serve this purpose.

```
epicalc::des(AgChem.table)

epicalc::label.var(ID,        "ID",
  dataFrame=AgChem.table)
epicalc::label.var(Formula,   "AgChem Formula",
  dataFrame=AgChem.table)
epicalc::label.var(TimeOfDay, "Time-of-Day",
  dataFrame=AgChem.table)
epicalc::label.var(Larvae,    "Larvae",
  dataFrame=AgChem.table)

epicalc::des(AgChem.table)
  # Confirm the description of each object variable.
```

Numeric codes have been used in this dataset to differentiate between breakout groups for the two factor-type object variables: AgChem.table$Formula and AgChem.table$TimeOfDay. Coerce objects into correct format.

```
# Object class before coercion

class(AgChem.table)
class(AgChem.table$ID)        # DataFrame$ObjectName notation.
class(AgChem.table$Formula)   # DataFrame$ObjectName notation.
```

```
class(AgChem.table$TimeOfDay) # DataFrame$ObjectName notation.
class(AgChem.table$Larvae)    # DataFrame$ObjectName notation.

# Coercion

AgChem.table$ID          <- as.factor(AgChem.table$ID)
AgChem.table$Formula     <- as.factor(AgChem.table$Formula)
AgChem.table$TimeOfDay   <- as.factor(AgChem.table$TimeOfDay)
AgChem.table$Larvae      <- as.numeric(AgChem.table$Larvae)

# Object class after coercion

class(AgChem.table)
class(AgChem.table$ID)         # DataFrame$ObjectName notation.
class(AgChem.table$Formula)    # DataFrame$ObjectName notation.
class(AgChem.table$TimeOfDay)  # DataFrame$ObjectName notation.
class(AgChem.table$Larvae)     # DataFrame$ObjectName notation.
```

Use the str() function to confirm object format. Note the details for str() output, especially the output against the data frame AgChem.table. Trust that actions have been implemented correctly, but verify too. The str() function should be used regularly to verify if data are in correct format.

```
str(AgChem.table)
str(AgChem.table$ID)           # DataFrame$ObjectName notation.
str(AgChem.table$Formula)      # DataFrame$ObjectName notation.
str(AgChem.table$TimeOfDay)    # DataFrame$ObjectName notation.
str(AgChem.table$Larvae)       # DataFrame$ObjectName notation.
```

Use the epicalc::des() function to again describe the data frame currently in use.

```
epicalc::des(AgChem.table)
```

Then, as another means of confirming that data are in correct format, use the levels() function against the factor object variables.

```
levels(AgChem.table$ID)
levels(AgChem.table$Formula)
levels(AgChem.table$TimeOfDay)
```

Use the summary() function against the object AgChem.table, which is a data frame, to gain an initial sense of descriptive statistics and frequency distributions. Other functions will expand on the summary() function, but this is always a good start for reviewing the data and conducting an additional quality assurance check.

```
summary(AgChem.table)
```

This initial review provides support that the dataset is in correct format. However, it is generally difficult to work with numeric values for the factor object variables Formula and TimeOfDay. Use the code book to review the meaning for each factor code and then note how this problem is easy to accommodate. There are more than a few ways to apply labels to factor object variables when using R. The labels() function is a fairly straight-forward way to achieve this aim.

```
# Apply the labels() function
AgChem.table$Formula.recode  <- factor(AgChem.table$Formula,
  labels = c("AgChem Formula 1",
             "AgChem Formula 2",
             "AgChem Formula 3"))

head(AgChem.table$Formula) # View initial data.
head(AgChem.table$Formula.recode)

summary(AgChem.table$Formula) # View descriptive statistics.
summary(AgChem.table$Formula.recode)

par(ask=TRUE)
epicalc::tab1(AgChem.table$Formula.recode,
  decimal=2,                        # Use the tab1() function
  sort.group=FALSE,                 # from the epicalc
  cum.percent=TRUE,                 # package to see details
  graph=TRUE,                       # about the selected
  missing=TRUE,                     # object variable. (The
  bar.values=c("frequency"),        # 1 of tab1 is the one
  horiz=FALSE,                      # numeric character and
  cex=1.15,                         # it is not the letter
  cex.names=1.15,                   # l).
  cex.lab=1.15,
  cex.axis=1.15,
  main="Factor Levels for Object Variable Formula.recode",
  ylab="Frequency of Formula.recode Factor Levels",
  col = c(rainbow(4)),
  gen=TRUE)
  # The number of colors in the rainbow() function was set
  # at one more than the number of levels for the factor
  # object variable, to better accommodate graphical
  # representation of NAs.

# Apply the labels() function
AgChem.table$TimeOfDay.recode <- factor(AgChem.table$TimeOfDay,
  labels = c("AM Application",
             "PM Application"))

head(AgChem.table$TimeOfDay) # View initial data.
head(AgChem.table$TimeOfDay.recode)

summary(AgChem.table$TimeOfDay)  # View descriptive statistics.
summary(AgChem.table$TimeOfDay.recode)

par(ask=TRUE)
epicalc::tab1(AgChem.table$TimeOfDay.recode,
  decimal=2,                        # Use the tab1() function
```

```
sort.group=FALSE,          # from the epicalc
cum.percent=TRUE,          # package to see details
graph=TRUE,                # about the selected
missing=TRUE,              # object variable. (The
bar.values=c("frequency"), # 1 of tab1 is the one
horiz=FALSE,               # numeric character and
cex=1.15,                  # it is not the letter
cex.names=1.15,            # l).
cex.lab=1.15,
cex.axis=1.15,
main="Factor Levels for Object Variable
TimeOfDay.recode",
ylab="Frequency of TimeOfDay.recode Factor Levels",
col = c(rainbow(3)),
gen=TRUE)
# The number of colors in the rainbow() function was set
# at one more than the number of levels for the factor
# object variable, to better accommodate graphical
# representation of NAs.
```

As a review, the object variable AgChem.table$Formula.recode was created by applying the factor() function against the object variable AgChem.table$Formula and the object variable AgChem.table $TimeOfDay.recode was created by applying the factor() function against the object variable AgChem.table$TimeOfDay. The labels() function was used to embellish future output into simple English. The head() function was used to view the first few lines of data, for quality assurance purposes. Other actions, such as use of the summary() function and the epicalc::tab1() function, provided additional review of intended actions as the original object variable was recoded into a new object variable.

If time permits, experiment with use of the factor() function and the as.factor() function. Would both functions yield the same result in the above example?

Now, merely use the attach() function again to confirm that all data are attached to the data frame.

```
attach(AgChem.table)
head(AgChem.table)
tail(AgChem.table)
summary(AgChem.table)  # Quality assurance data check.

str(AgChem.table)      # List all objects, with finite detail.
```

To continue with attention to quality assurance, use the table() function to see how data have been summarized using the newly created names (factor object variables) as well as the original names for the numeric object variables.

```
table(AgChem.table$Formula,        useNA = c("always"))
table(AgChem.table$Formula.recode, useNA = c("always"))
```

```
table(AgChem.table$TimeOfDay,          useNA = c("always"))
table(AgChem.table$TimeOfDay.recode, useNA = c("always"))

table(AgChem.table$Formula,
      AgChem.table$TimeOfDay, useNA = c("always"))
table(AgChem.table$Formula.recode,
      AgChem.table$TimeOfDay.recode, useNA = c("always"))
```

Observe how the argument useNA = c("always") is used with the table function, to force identification of missing values.

To a student or beginning researcher, this type of redundancy and attention to detail at this stage of development may seem unnecessary, but it more than helps reduce later errors caused by a simple oversight. Quality assurance needs to be part of a continuous process.

4.3   Conduct a Visual Data Check

Although functions such as summary(), min(), and max(), are used to review data, a visual data check process is also quite useful and should always be part of the initial quality assurance process. Simple plots and bar charts, initially, may be all that is needed to look for data that are either illogical or out-of-range. More complex graphical images may be generated later, but for now attention should be directed to the use of simple visual images for quality purposes.

```
names(AgChem.table)      # Confirm all object variables.
```

4.3.1   Simple Plots

Simple plots are often more than sufficient, at the beginning of a project, to gain a sense of trends. More complex plots are needed, but only after initial efforts are completed.

```
par(ask=TRUE)
plot(AgChem.table$ID,
  main="AgChem.table$ID Visual Data Check",
  col = c(rainbow(300)))
  # Factor object variable breakout groups.

par(ask=TRUE)
plot(AgChem.table$Formula,
  main="AgChem.table$Formula Visual Data Check",
  col = c(rainbow(3)))
  # Factor object variable breakout groups.
```

```
par(ask=TRUE)
plot(AgChem.table$Formula.recode,
  main="AgChem.table$Formula.recode Visual Data Check",
  col = c(rainbow(3)))
  # Factor object variable breakout groups.

par(ask=TRUE)
plot(AgChem.table$TimeOfDay,
  main="AgChem.table$TimeOfDay Visual Data Check",
  col = c(rainbow(2)))
  # Factor object variable breakout groups.

par(ask=TRUE)
plot(AgChem.table$TimeOfDay.recode,
  main="AgChem.table$TimeOfDay.recode Visual Data Check",
  col = c(rainbow(4)))
  # Factor object variable breakout groups.

par(ask=TRUE)
plot(AgChem.table$Larvae,
  main="AgChem.table$Larvae Visual Data Check",
  pch=19,
  col = c("black"))
  # Note how the datapoints are represented as
  # black solid circles for this numeric object
  # variable.
```

These simple plots offer a brief view of the data and they are also useful in an attempt to identify the possible presence of outliers and any data that may seem illogical or out-of-range. In an attempt to look for outliers, the ylim argument has been avoided, so that all data are plotted. Is an extreme value an incorrect datum that needs correction, due to a human error (e.g., measurement, data input, etc.) during some part of the data life cycle? Or, is an extreme value an outlier that calls for a reexamination of assumptions and possible redirection for future research? Initial plots help experienced researchers address these concerns.

4.3.2   Histogram of the Summary Object Variable

This sample lesson has been designed to look into the nature of object variable Larvae (the number of larvae of a specific but unidentified insect, counted in a one square meter area, with the area subject to random selection), and the factor object variables Formula (three separate formulations of an agricultural chemical) and TimeOfDay (the time-of-day the agricultural chemical was applied, either in the early morning (AM) or in the late afternoon (PM)). The factor object variables were recoded into a more verbose format as object variables `Formula.recode` and `TimeOfDay.recode`. Given the nature of Larvae counts, it is best to supplement the `plot()` function with other functions, to gain a different view of the continuous values of larvae, overall and by breakout groups.

```
# Simple histogram
par(ask=TRUE)
hist(AgChem.table$Larvae,
   main="AgChem.table$Larvae Visual Data Check (Histogram)",
   font=2, cex.lab=1.15, col="red")

# A histogram with attention to axis values
par(ask=TRUE)
hist(AgChem.table$Larvae,
   main="Histogram of Larvae Counts",
   xlab="Larvae Counts (Limit = 0 to 180)",
   ylab="Frequency",
   xlim=c(100,180),     # Note the selection for xlim.
   ylim=c(0,220),       # Note the selection for ylim.
   cex.lab=1.15, cex.axis=1.15, freq=TRUE,
   border="blue", col="red")
   # Note the ylim=c(0,220) argument.  The Y axis is pushed
   # out to 220 to extend the axis presentation.
```

The histogram is a simple, but useful graphical tool to learn more about the distribution of values for an object variable. Note below how the histogram can be embellished to show additional information.

```
# Histogram with frequency values above column bars
par(ask=TRUE)
hist(AgChem.table$Larvae,
   main="Histogram of Larvae Counts",
   xlab="Larvae Counts (Limit = 0 to 180)",
   ylab="Frequency",
   xlim=c(100,180),     # Note the selection for xlim.
   ylim=c(0,220),       # Note the selection for ylim.
   breaks="Sturges",
   font=2, font.main=2, font.lab=2, lwd=2, freq=TRUE,
   include.lowest=TRUE, right=TRUE, axes=TRUE,
   plot=TRUE, labels=TRUE, border="blue", col="red")
box()

# Histogram with the rug() function used on the X axis
par(ask=TRUE)
hist(AgChem.table$Larvae,
   main="Histogram of Larvae Counts,
```

```
With an Added Rug",
  xlab="Larvae Counts (Limit = 0 to 180)",
  xlim=c(120,160),  # Scale adjusted for presentation.
  ylim=c(0,50),      # Scale adjusted for presentation.
  font.lab=2, font.axis=2, border="blue", col="red",
  nclass=50, freq=TRUE, prob=FALSE)
rug(AgChem.table$Larvae,  # Look at the bottom line.
  col='darkblue',
  ticksize = 0.05,
  side=1,
  lwd=2)
box()

help(fivenum)
fivenum(AgChem.table$Larvae)

# Histogram with fivenum() output added to a legend
par(ask=TRUE)
hist(AgChem.table$Larvae,
  main="Histogram of Larvae Counts, With fivenum()
  Output:  Minimum, 1Q, Median,
  3Q, Maximum",
  xlab="Larvae Counts (Limit = 0 to 180)",
  xlim=c(120,160),  # Scale adjusted for presentation.
  ylim=c(0,50),      # Scale adjusted for presentation.
  font.lab=2, font.axis=2, border="blue", col="red",
  nclass=50, freq=TRUE, prob=FALSE)
  axis(side=1,
    line=-0.1,
    at=fivenum(AgChem.table$Larvae)[1:5],
    lwd=1,
    tick=TRUE,
    font=2,
    labels=c("Minimum", "1Q","Median","3Q", "Maximum"),
    cex.axis=1.05,
    col.axis="black",
    padj=-2.9)
savefamily   <- par(family="mono") # Courier font
savefont     <- par(font=2)        # Bold
legend("topright",
  legend = c(
  "Tukey's five number summary",
  "=================================",
  "> fivenum(AgChem.table$Larvae)     ",
  "[1] 122.0 135.5 138.0 140.0 156.0 ",
  "---------------------------------",
```

```
"122.0  Minimum                        ",
"135.5  Lower-Hinge (1Q)               ",
"138.0  Median                         ",
"140.0  Upper-Hinge (3Q)               ",
"156.0  Maximum                        ",
"----------------------------------"),
  ncol=1,
  locator(1),
  xjust=1,
  text.col="black",
  cex=1.15,
  inset=0.025,
  bty="n")
par(savefamily)
par(savefont)
  # The legend for this figure is rather large.
  # It was set to a Courier-type mono-spaced font
  # to have all fivenum() function values and
  # character values line up properly.
```

Histograms can be even more complex, if needed. The following graphic may be somewhat redundant, but it will help reinforce attention to normal distribution. If this is a concern, R can easily accommodate nonparametric tests that are not as dependent on normal distribution. For now, however, focus on distribution patterns to see if the examined object variable supports what some may claim are the more robust parametric tests, such as Two-Way ANOVA.

```
# Create the object variable Larvae.mean
Larvae.mean <- mean(AgChem.table$Larvae,
  na.rm=TRUE) # Note the need for na.rm=TRUE.
Larvae.mean

# Create the object variable Larvae.sd
Larvae.sd    <- sd(AgChem.table$Larvae,
  na.rm=TRUE) # Note the need for na.rm=TRUE.
Larvae.sd

# Histogram with normal curve overlay
par(ask=TRUE)
hist(AgChem.table$Larvae,
  main="Histogram of Larvae Counts",
  xlab="Larvae Counts (Limit = 0 to 180)",
  xlim=c(100,180),    # Note the selection for xlim.
  ylim=c(0,0.15),     # Note the selection for ylim.
  font.lab=2, font.axis=2, border="blue", col="red",
  nclass=15, prob=TRUE)
```

```
curve(dnorm(x, mean=Larvae.mean, sd=Larvae.sd),
   col="darkblue",
   lwd=2,
   add=TRUE)
box()
```

Notice how xlim=c(100,180) and ylim=c(0,0.15) were used to set the limits for the *X*-axis and the *Y*-axis. The *X*-axis was purposely set to go to 180 even though the maximum value of Larvae is 156. This action extended the normal curve overlay to the right and allowed presentation of the entire curve. In the same manner, the *Y*-axis was set to allow presentation of the upper limits of the histogram.

On this topic, it takes practice, trial, error, and some degree of judgment to come up with the best selection for ylim when using graphics. Try a few different selections to see the scale that meets requirements for viewing at a distance, assuming the graphic may be presented in a large lecture hall with outside light coming in from windows and open doors.

The density plot overlay provides another view of how values for the object variable are distributed, allowing for even greater insight into the nature of the data. Again, notice how settings for the *X*-axis and the *Y*-axis are purposely set to support a complete display of the histogram and the density plot overlay.

```
# Histogram with density plot overlay
par(ask=TRUE)
hist(AgChem.table$Larvae,
   main="Histogram of Larvae Counts",
   xlab="Larvae Counts (Limit = 0 to 180)",
   xlim=c(100,180),     # Note the selection for xlim.
   ylim=c(0,0.15),      # Note the selection for ylim.
   font.lab=2, font.axis=2, border="blue", col="red",
   nclass=15, prob=TRUE)
lines(density(AgChem.table$Larvae,
   na.rm=TRUE),         # Note the need for na.rm=TRUE.
   col="darkblue",
   lwd=2)
box()
```

Notice how the label for the *Y*-axis reads density and that it does not read frequency. Do not confuse a density plot with the normal curve.

This density plot is interesting in that there seems to be two unique distributions here. A decision may be necessary to determine if the data support two-way ANOVA, as presented in these lessons, or if the nonparametric Friedman two-way ANOVA by Ranks test is the more appropriate choice. Generally, this is more a matter of knowledge about statistics and not necessarily a matter of whether R or some other tool is used to conduct the analysis.

4.3.3 Horizontal and Vertical Boxplots of the Summary Object Variable

The boxplot is often seen in scientific literature and Web pages detailing analyses for corporate stocks and bonds. It is one of a growing collection of visual tools used to represent data distribution. Largely depending on preference and need, a boxplot can be displayed in either horizontal or vertical orientation while still highlighting the lower quartile (Q1), the median (Q2), the upper quartile (Q3), and any possible outliers. If a boxplot were selected for a large group presentation, it may be necessary to provide details on the nature of a boxplot. The general public may have some background with histograms and the normal curve, but few members of the general public know about the boxplot and they usually have trouble understanding it and what it means.

In the following set of graphics, the boxplot is embellished in many ways: the range is expanded to mask outliers, an overlay chart is added, and a stripchart is embedded over the boxplot. R allows for this type of flexibility, but good judgment should always be used before these additional tools are used. Consider the purpose of the image, the needs of the receiving audience, the medium, etc. The purpose of a graphical image, such as a boxplot, is to communicate with the reader relationships and association between and among the data. Overly busy images run the risk of confusing the reader, limiting the advantage of using a graphical image. Prudence is always the best practice when preparing graphical images.

```
# Horizontal boxplot
par(ask=TRUE)
boxplot(AgChem.table$Larvae,
   horizontal=TRUE,
   main="Horizontal Boxplot of Larvae Counts",
   xlab="Larvae Counts (Limit = 0 to 180)",
   ylim=c(100,180),  # Note the selection for ylim.
   cex.lab=1.15,
   cex.axis=1.15,
   lty=1,             # Note the line type.
   lwd=3,             # Note the line width.
   border="blue",
   col="red")
box()
   # Dots serve as outliers.

# Vertical boxplot
par(ask=TRUE)
boxplot(AgChem.table$Larvae,
   horizontal=FALSE,
   main="Vertical Boxplot of Larvae Counts",
   ylab="Larvae Counts (Limit = 0 to 180)",
   ylim=c(100,180),  # Note the selection for ylim.
```

```
  cex.lab=1.15,
  cex.axis=1.15,
  lty=1,               # Note the line type.
  lwd=3,               # Note the line width.
  border="blue",
  col="red")
box()
  # Dots serve as outliers.

# Horizontal boxplot without outliers
par(ask=TRUE)
boxplot(AgChem.table$Larvae,
  horizontal=TRUE,
  main="Horizontal Boxplot of Larvae Counts,
  With Range Set to 0 to Mask Outliers",
  xlab="Larvae Counts (Limit = 0 to 180)",
  ylim=c(100,180),   # Note the selection for ylim.
  cex.lab=1.15,
  cex.axis=1.15,
  lty=1,               # Note the line type.
  lwd=3,               # Note the line width.
  border="blue",
  col="red",
  range=0)             # Note the value for range.
box()
  # In this boxplot, the range=0 argument has been
  # used to show the full range for the distribution
  # of all data.  The outliers do not show.  It is
  # common to set the boxplot so that the outliers
  # show, but there may be times when it is useful to
  # use range=0 so that the outliers do not show.

# Vertical boxplot with an overlay grid
par(ask=TRUE)
boxplot(AgChem.table$Larvae,
  horizontal=FALSE,
  main="Vertical Boxplot of Larvae Counts,
  With Range Set to 0 to Mask Outliers,
  and an Overlay Grid",
  xlab="Larvae Counts (Limit = 0 to 180)",
  ylim=c(100,180),   # Note the selection for ylim.
  cex.lab=1.15,
  cex.axis=1.15,
  lty=1,               # Note the line type.
  lwd=3,               # Note the line width.
```

```
  border="blue",
  col="red",
  range=0); grid(col="gray", lwd=1, lty="dashed")
box()
  # The grid() function was used, to overlay a grid on
  # top of the graphical image.

# Horizontal boxplot with an overlay stripchart
par(ask=TRUE)
boxplot(AgChem.table$Larvae,
  horizontal=TRUE,
  main="Horizontal Boxplot of Larvae Counts,
  With Range Set to 0 to Mask Outliers,
  an Overlay Grid, and a Stripchart",
  xlab="Larvae Counts (Limit = 0 to 180)",
  ylim=c(100,180),   # Note the selection for ylim.
  cex.lab=1.15,
  cex.axis=1.15,
  lty=1,             # Note the line type.
  lwd=3,             # Note the line width.
  border="blue",
  col="red",
  range=0); grid(col="gray", lwd=1, lty="dashed")
stripchart(AgChem.table$Larvae,
  method="jitter",
  jitter=.05,
  pch=1,
  vertical=FALSE,
  add=TRUE)
box()
  # The stripchart() function was used to overlay a
  # stripchart on the boxplot.  Notice how boxplot()
  # horizontal=TRUE and stripchart() vertical=FALSE.
```

4.3.4   Violin Plot of the Summary Object Variable

The violin plot is a fairly new, but increasingly useful, tool in the growing array of graphical tools used to represent data and specifically data distribution. A typical violin plot incorporates features from the boxplot (such as representation of the median and the interquartile range) and the kernel density plot.

```
install.packages("vioplot")
library(vioplot)        # Load the vioplot package.
help(package=vioplot)   # Show the information page.
```

```
sessionInfo()                    # Confirm all attached
                                   packages.

# Horizontal violin plot
par(ask=TRUE)
savelab.cex  <- par(cex.lab=1.15)
saveaxis.cex <- par(cex.axis=1.15)
vioplot::vioplot(AgChem.table$Larvae,
  horizontal=TRUE,
  names=c("Larvae Counts (0 to 180)"),
  lty=1,
  lwd=6,
  col="cyan",
  rectCol="blue",   # Mark the interquartile range.
  colMed="red",     # Mark the median.
  pchMed=19,
  border="black",
  drawRect=TRUE,
  ylim=c(100,180))
  title("Horizontal Violin Plot of Larvae Counts")
par(savelab.cex)  # Go back to default setting.
par(saveaxis.cex) # Go back to default setting.

# Vertical violin plot
par(ask=TRUE)
savelab.cex  <- par(cex.lab=1.15)
saveaxis.cex <- par(cex.axis=1.15)
vioplot::vioplot(AgChem.table$Larvae,
  horizontal=FALSE,
  names=c("Larvae Counts (0 to 180)"),
  lty=1,
  lwd=6,
  col="cyan",
  rectCol="blue",   # Mark the interquartile range.
  colMed="red",     # Mark the median.
  pchMed=19,
  border="black",
  drawRect=TRUE,
  ylim=c(100,180))
  title("Vertical Violin Plot of Larvae Counts")
par(savelab.cex)  # Go back to default setting.
par(saveaxis.cex) # Go back to default setting.
```

4.3.5   Beanplot of the Summary Object Variable

The beanplot provides another view of how the distribution of data can be presented. A careful review of the beanplot will show how it includes features found in the boxplot, stripchart, and the violin plot while offering a few unique features.

```
install.packages("beanplot")
library(beanplot)        # Load the beanplot package.
help(package=beanplot)   # Show the information page.
sessionInfo()            # Confirm all attached packages.

# Horizontal beanplot
par(ask=TRUE)
beanplot::beanplot(AgChem.table$Larvae,
  main="Horizontal Beanplot of Larvae Counts,
  Beans do not Show",
  xlab="Larvae Count",
  ll=0,   # Hide the beans.
  cex.lab=1.25,
  cex.axis=1.25,
  ylim=c(120,160),
  method="stack",
  overallline="mean",
  horizontal=TRUE,
  col="red",
  border = "blue",
  show.names=TRUE,
  log="")
legend("topleft",
  legend = c("Mean 139.6 and SD 7.0"),
    ncol=1,
    locator(1),
    xjust=1,
    text.col="black",
    cex=1.25,
    inset=0.05,
    bty="n")

# Vertical beanplot
par(ask=TRUE)
beanplot::beanplot(AgChem.table$Larvae,
  main="Vertical Beanplot of Larvae Counts,
  Beans do not Show",
  xlab="Larvae Count",
  ll=0,   # Hide the beans.
```

```
cex.lab=1.25,
cex.axis=1.25,
ylim=c(120,160),
method="stack",
overallline="mean",
horizontal=FALSE,
col="red",
border = "blue",
show.names=TRUE,
log="")
legend("topleft",
  legend = c("Mean 139.6 and SD 7.0"),
    ncol=1,
    locator(1),
    xjust=1,
    text.col="black",
    cex=1.25,
    inset=0.05,
    bty="n")
```

As demonstrated in these beanplots, look at the unusual distribution pattern
for the data and how there is possibly at least one group of data that is quite
different from the rest. This oddity, if indeed this is the right term, clearly shows
in these beanplots. Immediately, this observation is interesting and it will give
justification for detailed empirical analyses, after these exploratory graphical figures
are generated and reviewed. This initial knowledge about the data would be lost
if these graphical tools, serving as advance organizers, were given only minimal
attention and there were a rush to go immediately to numerical analyses.

4.3.6 Quantile–Quantile (Q–Q) Plot of the Summary Object Variable

Although perhaps used less frequently than other graphical tools, the Quantile–
Quantile (Q–Q) plot is also used to gain a full understanding of the summary object
variable. The Q–Q plot is similar to a probability plot and it is particularly useful to
view distributions of values and behavior of the tails.

```
# Q-Q plot
par(ask=TRUE)
qqnorm(AgChem.table$Larvae,
  main="Q-Q Plot of Larvae Counts",
  ylim=c(0,180),       # Note the selection for ylim.
  cex.lab=1.15,
  cex.axis=1.15,
  font.lab=2,
```

```
font.axis=2,
plot.it=TRUE,
pch=19,
col="red")
box()
```

In the same way that descriptive statistics at the breakout level provide detail about selected object variables for specific groups (e.g., pounds of milk per lactation for Holstein dairy cattle versus pounds of milk per lactation for Jersey dairy cattle), graphical displays at the breakout level are also useful. It is especially helpful to graphically display object variables at the breakout level to compare distribution, extreme values, etc. This is all done in an effort to determine, later, if there are (or are not) differences by breakout groups.

Imagine that a Nebraska farmer has more than 1,000 acres of wheat, represented by five different wheat varieties. The farmer is certainly interested in total wheat yield per acre, but the farmer is also keenly interested in wheat yield per acre for each wheat variety. Better yet, the farmer is even more interested in knowing wheat yield per acre by wheat variety and by soil type assuming there are three dominant soil types in the growing area. By knowing wheat yield per acre by variety, the farmer can make an informed decision on management practices. By knowing wheat yield per acre by variety and by soil type the farmer can make an even better informed decision on what to plant next year, where, in an effort to optimize yields in the future. Graphical displays at the breakout level provide this type of granularity.

4.3.7   Sorted Dot Chart of the Summary Object Variable by Breakout Object Variables

```
par(ask=TRUE) #Formula.recode
epicalc::summ(AgChem.table$Larvae,
  by=AgChem.table$Formula.recode,
  graph=TRUE, # Use graph=TRUE, if desired.
  pch=20,
  ylab="auto",
  main="Sorted Dotplot of Larvae Counts
  by AgChem Formula",
  cex.X.axis=1.10, # Note X axis label size.
  cex.Y.axis=1.10, # Note Y axis label size.
  font.lab=2,
  dot.col="auto")
  # Give special attention to the graphical
  # output for AgChem Formula 1.

par(ask=TRUE) # TimeOfDay.recode
epicalc::summ(AgChem.table$Larvae,
```

```
by=AgChem.table$TimeOfDay.recode,
graph=TRUE, # Use graph=TRUE, if desired.
pch=20,
ylab="auto",
main="Sorted Dotplot of Larvae Counts
by AgChem Application Time-of-Day",
cex.X.axis=1.10, # Note X axis label size.
cex.Y.axis=1.10, # Note Y axis label size.
font.lab=2,
dot.col="auto")
# Give special attention to the graphical
# output for PM Application.
```

These sorted dot charts provide a clear view of larvae counts by AgChem formula and then by time-of-day. The epicalc::summ() function and the accompanying graph=TRUE argument should always be among the first graphical tools used to examine distributions by breakout group.

4.3.8 Histogram of the Summary Object Variable by Breakout Object Variables

It is certainly possible to put all larvae counts for AgChem formula 1 into a separate object and to then prepare a histogram of only AgChem formula 1 counts, either by using R or a spreadsheet. This action could then be repeated so that there are separate object variables for all AgChem formula breakout groups: AgChem formula 1, AgChem formula 2, and AgChem formula 3. By using the lattice package, however, this excessively redundant action can be avoided, with separate histograms for each AgChem formula breakout group placed into the same graphical image. This action can then be repeated for the two application time-of-day breakout groups: AM application and PM application. This general approach and subsequent use of the lattice package can be used for other graphic techniques, as needed.

4.3.8.1 Breakout Histograms by Count

```
install.packages("lattice")
library(lattice)          # Load the lattice package.
help(package=lattice)     # Show the information page.
sessionInfo()             # Confirm all attached packages.

par(ask=TRUE) # Formula.recode 1 Column by 3 Rows
lattice::histogram(~AgChem.table$Larvae |
                  AgChem.table$Formula.recode,
```

```
type="count",        # Note:   count
par.settings=simpleTheme(lwd=2),
par.strip.text=list(cex=1.15, font=2),
scales=list(cex=1.15),
main="Histograms (Count) of Larvae Counts
by AgChem Formula",
xlab=list("Larvae Counts (Limit = 0 to 180)",
cex=1.15, font=2),
xlim=c(0,180),       # Note the range.
ylab=list("Count", cex=1.15, font=2),
aspect=0.2,
layout = c(1,3),     # Note:   1 Column by 3 Rows.
col="red")

par(ask=TRUE) # Formula.recode 3 Columns by 1 Row
lattice::histogram(~AgChem.table$Larvae |
                   AgChem.table$Formula.recode,
type="count",        # Note:   count
par.settings=simpleTheme(lwd=2),
par.strip.text=list(cex=1.15, font=2),
scales=list(cex=1.15),
main="Histograms (Count) of Larvae Counts
by AgChem Formula",
xlab=list("Larvae Counts (Limit = 0 to 180)",
cex=1.15, font=2),
xlim=c(100,180),     # Note the range.
ylab=list("Count", cex=1.15, font=2),
aspect=0.2,
layout = c(3,1),     # Note:   3 Columns by 1 Row.
col="red")

par(ask=TRUE) # TimeOfDay.recode 1 Column by 2 Rows
lattice::histogram(~AgChem.table$Larvae |
                   AgChem.table$TimeOfDay.recode,
type="count",        # Note:   count
par.settings=simpleTheme(lwd=2),
par.strip.text=list(cex=1.15, font=2),
scales=list(cex=1.15),
main="Histograms (Count) of Larvae Counts
by Time-Of-Day",
xlab=list("Larvae Counts (Limit = 0 to 180)",
cex=1.15, font=2),
xlim=c(0,180),       # Note the range.
ylab=list("Count", cex=1.15, font=2),
aspect=0.2,
```

```
          layout = c(1,2),   # Note:  1 Column by 2 Rows.
          col="red")

par(ask=TRUE) # TimeOfDay.recode 2 Columns by 1 Rows
lattice::histogram(~AgChem.table$Larvae |
                        AgChem.table$TimeOfDay.recode,
          type="count",      # Note:  count
          par.settings=simpleTheme(lwd=2),
          par.strip.text=list(cex=1.15, font=2),
          scales=list(cex=1.15),
          main="Histograms (Count) of Larvae Counts
          by Time-Of-Day",
          xlab=list("Larvae Counts (Limit = 0 to 180)",
          cex=1.15, font=2),
          xlim=c(100,180),   # Note the range.
          ylab=list("Count", cex=1.15, font=2),
          aspect=0.2,
          layout = c(2,1),   # Note:  2 Columns by 1 Rows.
          col="red")
```

Look at the different scales used in this set of figures. In the first figure the scale is xlim=c(0,180) and in the next figure the scale is xlim=c(100,180). Experiment with different arrangements for columns and rows and experiment with different scales, to see the best way to present outcomes in graphical format.

4.3.8.2   Breakout Histograms by Percent

```
par(ask=TRUE) # Formula.recode
lattice::histogram(~AgChem.table$Larvae |
                        AgChem.table$Formula.recode,
          type="percent",    # Note:  percent
          par.settings=simpleTheme(lwd=2),
          par.strip.text=list(cex=1.15, font=2),
          scales=list(cex=1.15),
          main="Histograms (Percent) of Larvae Counts
          by AgChem Formula",
          xlab=list("Larvae Counts (Limit = 0 to 180)",
          cex=1.15, font=2),
          xlim=c(0,180),     # Note the range.
          ylab=list("Percent", cex=1.15, font=2),
          aspect=0.2,
          layout = c(1,3),   # Note:  1 Column by 3 Rows.
          col="red")
```

```
par(ask=TRUE) # Formula.recode
lattice::histogram(~AgChem.table$Larvae |
                    AgChem.table$Formula.recode,
  type="percent",    # Note:  percent
  par.settings=simpleTheme(lwd=2),
  par.strip.text=list(cex=1.15, font=2),
  scales=list(cex=1.15),
  main="Histograms (Percent) of Larvae Counts
  by AgChem Formula",
  xlab=list("Larvae Counts (Limit = 0 to 180)",
  cex=1.15, font=2),
  xlim=c(100,180),   # Note the range.
  ylab=list("Percent", cex=1.15, font=2),
  aspect=0.2,
  layout = c(3,1),   # Note:  3 Columns by 1 Row.
  col="red")

par(ask=TRUE) # TimeOfDay.recode
lattice::histogram(~AgChem.table$Larvae |
                    AgChem.table$TimeOfDay.recode,
  type="percent",    # Note:  percent
  par.settings=simpleTheme(lwd=2),
  par.strip.text=list(cex=1.15, font=2),
  scales=list(cex=1.15),
  main="Histograms (Percent) of Larvae Counts
  by Time-Of-Day",
  xlab=list("Larvae Counts (Limit = 0 to 180)",
  cex=1.15, font=2),
  xlim=c(0,180),     # Note the range.
  ylab=list("Percent", cex=1.15, font=2),
  aspect=0.2,
  layout = c(1,2),   # Note:  1 Column by 2 Rows.
  col="red")

par(ask=TRUE) # TimeOfDay.recode
lattice::histogram(~AgChem.table$Larvae |
                    AgChem.table$TimeOfDay.recode,
  type="percent",    # Note:  percent
  par.settings=simpleTheme(lwd=2),
  par.strip.text=list(cex=1.15, font=2),
  scales=list(cex=1.15),
  main="Histograms (Percent) of Larvae Counts
  by Time-Of-Day",
  xlab=list("Larvae Counts (Limit = 0 to 180)",
  cex=1.15, font=2),
```

```
xlim=c(100,180),  # Note the range.
ylab=list("Percent", cex=1.15, font=2),
aspect=0.2,
layout = c(2,1),  # Note:  2 Columns by 1 Row.
col="red")
```

The lattice::histogram() function, using both count and percent, is a
valuable tool in an attempt to make informative but condensed graphical presentations. The type="count" argument may be a concern when there is a fairly large
unequal N for the breakout groups, but the unequal N for breakout groups is not a
major concern when the type="percent" argument is used. The histograms
based on percent give a sense of the distribution pattern for all three AgChem
formula groups and both Time-Of-Day groups. Whether these figures are ever
published, these distribution patterns should always be confirmed before the use of
any inferential statistical tests. Quality assurance is a constant process.

4.3.9   Density Plot of the Summary Object Variable by Breakout Object Variables

The lattice::densityplot() function is used to prepare a fairly simple
presentation of density plots for the breakout object variable. The density plot
provides another view of data and more importantly, data distribution.

```
par(ask=TRUE)
lattice::densityplot(~AgChem.table$Larvae |
                        AgChem.table$Formula.recode,
  par.settings = simpleTheme(lwd=2),
  par.strip.text=list(cex=1.15, font=2),
  scales=list(cex=1.15),
  main="Density Plot of Larvae Counts by AgChem Formula",
  xlab=list("Larvae Counts (Limit = 0 to 180)",
  cex=1.15, font=2),
  xlim=c(100,180),       # The scale improves presentation.
  ylab=list("Density", cex=1.15, font=2),
  aspect=0.2,
  layout = c(1,3),  # Note:  1 Columns by 3 Rows.
  lwd=6,        # A thick line improves presentation.
  col="darkred")

par(ask=TRUE)
lattice::densityplot(~AgChem.table$Larvae |
                        AgChem.table$TimeOfDay.recode,
  par.settings = simpleTheme(lwd=2),
  par.strip.text=list(cex=1.15, font=2),
  scales=list(cex=1.15),
```

```
main="Density Plot of Larvae Counts by Time-Of-Day",
xlab=list("Larvae Counts (Limit = 0 to 180)",
cex=1.15, font=2),
xlim=c(100,180),        # The scale improves presentation.
ylab=list("Density", cex=1.15, font=2),
aspect=0.2,
layout = c(1,2),  # Note:  1 Columns by 2 Rows
lwd=6,        # A thick line improves presentation.
col="darkred")
install.packages("sm")
library(sm)             # Load the sm package.
help(package=sm)        # Show the information page.
sessionInfo()           # Confirm all attached packages.
```

The sm::sm.density.compare() function is also used to graphically demonstrate density plots for the breakout object variable.

```
par(ask=TRUE)
saveline.width <- par(lwd=3) # Generate a heavy line
sm::sm.density.compare(AgChem.table$Larvae,
                       AgChem.table$Formula.recode,
  xlab=list("Larvae Counts (Limit = 0 to 180)",
  cex=1.15,
  font=2),
  ylab=list("Density", cex=1.15, font=2),
  xlim=c(100,180),       # The scale improves presentation.
  ylim=c(0,0.15))
title(main="Density Plot of Larvae Counts by AgChem Formula")
colorfill <- c(2:(2+length(levels(
  AgChem.table$Formula.recode))))
legend(locator(1), levels(AgChem.table$Formula.recode),
  fill=colorfill)
  # Click on an open location to paste the legend.
par(saveline.width) # Return to original value.

par(ask=TRUE)
saveline.width <- par(lwd=3) # Generate a heavy line
sm::sm.density.compare(AgChem.table$Larvae,
                       AgChem.table$TimeOfDay.recode,
  xlab=list("Larvae Readings (Limit = 0 to 180)",
  cex=1.15,
  font=2),
  ylab=list("Density", cex=1.15, font=2),
  xlim=c(100,180),
  ylim=c(0,0.15))
title(main="Density Plot of Larvae Counts by Time-Of-Day")
colorfill <- c(2:(2+length(levels(
  AgChem.table$TimeOfDay.recode))))
```

```
legend(locator(1), levels(AgChem.table$TimeOfDay.recode),
  fill=colorfill)
  # Click on an open location to paste the legend.
par(saveline.width) # Return to original value.
```

An interesting addition to these figures involves placement of the legend. Here, the legend is color coded and the legend is generated by clicking the mouse in any desired spot in the graphic. This is a simple demonstration of another way R can be used to embellish a graphical image.

4.3.10   Boxplot of the Summary Object Variable by Breakout Object Variables

```
par(ask=TRUE) # Note breakout group by measured object.
lattice::bwplot(AgChem.table$Formula.recode ~
                AgChem.table$Larvae,
  par.settings = simpleTheme(lwd=2),
  par.strip.text=list(cex=1.15, font=2),
  scales=list(cex=1.15),
  main="Boxplot of Larvae Counts by AgChem Formula",
  xlab=list("Larvae Counts (Limit = 0 to 180)",
  cex=1.15, font=2),
  xlim=c(100,180),
  ylab=list("AgChem Formula", cex=1.15, font=2),
  aspect=0.5,
  layout=c(1,1),
  col="red")

par(ask=TRUE) # Note breakout group by measured object.
lattice::bwplot(AgChem.table$TimeOfDay.recode ~
                AgChem.table$Larvae,
  par.settings = simpleTheme(lwd=2),
  par.strip.text=list(cex=1.15, font=2),
  scales=list(cex=1.15),
  main="Boxplot of Larvae Counts by Time-Of-Day",
  xlab=list("Larvae Counts (Limit = 0 to 180)",
  cex=1.15, font=2),
  xlim=c(100,180),
  ylab=list("Time-Of-Day", cex=1.15, font=2),
  aspect=0.5,
  layout=c(1,1),
  col="red")

par(ask=TRUE)
lattice::bwplot(~ Larvae | Formula.recode * TimeOfDay.recode,
  data = AgChem.table,
```

```
par.settings = simpleTheme(lwd=2),
par.strip.text=list(cex=1.15, font=2),
scales=list(cex=1.15),
main="Two-Way ANOVA:  Larvae ~ Formula + TimeOfDay\n
AgChem Formula = 3 Groups and Time-Of-Day = 2 Groups",
xlab=list("Larvae Count (Limit = 0 to 180)\n
AgChem Formula:
1 = AgChem 1, 2 = AgChem 2, 3 = AgChem 3",
cex=1.15, font=2),
xlim=c(100,180),
ylab=list("Time-Of-Day:
1 = AM, 2 = PM",
cex=1.15, font=2),
aspect=0.25,
layout=c(3,2),
col="red")
# The \n character sequence is used to force a
# new line, which improves presentation.
```

Look at the way outliers are identified, if there are any, as the small circles extending beyond the hinges of the boxplots. Outliers are always of interest and demand attention. Is an outlier a possible error in data entry that requires correction? Or, is the outlier indeed a correct datum? If so, how is it possible for individual values to deviate so much from norm values? Imagine if the number of larvae in one count were 250 instead of the more expected count of about 140 (Larvae Mean $= 139.6$ and SD $= 7.0$). This count is certainly possible, but follow-up actions would be prudent to be certain that this count is correct (as an outlier) and that it is not the result of an error in data entry. If the count of 250 were correct then this initiates a new set of actions, to look into how such an unexpected count could occur. Experienced researchers follow-up on unexpected outcomes.

When using the `lattice::bwplot()` function, as compared to other uses of the lattice package, the object variable representing breakout groups comes before the object variable representing the measured datum. Notice also that the two object variables are separated by the tilde (e.g., ~) character.

4.3.11   Horizontal and Vertical Boxplots of the Summary Object Variable by Breakout Object Variables

Although the `lattice::bwplot()` function may be more than sufficient, the production of both horizontal and vertical boxplots by breakout groups can also be achieved by slight modification to the R syntax previously used in this lesson. It is largely a personal preference to display a boxplot in either horizontal mode or vertical mode. To meet the needs of various readers, it is common to use both orientations. Consider both horizontal boxplots and vertical boxplots, to see which

alignment shows best, both in print as well as display in a large lecture hall using
some type of screen projection.

```
par(ask=TRUE)
boxplot(AgChem.table$Larvae ~ AgChem.table$Formula.recode,
  horizontal=TRUE, # Horizontal boxes in this case.
  main="Horizontal Boxplot of Larvae Counts
  by AgChem Formula",
  ylim=c(100,180),
  xlab="Larvae Count (Limit = 0 to 180)",
  ylab="AgChem Formula",
  cex.lab=1.15, cex.axis=0.95, las=3, border="blue",
  col="red")
box()


par(ask=TRUE)
boxplot(AgChem.table$Larvae ~ AgChem.table$Formula.recode,
  horizontal=FALSE,# Vertical boxes in this case.
  main="Vertical Boxplot of Larvae Counts
  by AgChem Formula",
  ylim=c(100,180),
  ylab="Larvae Counts (Limit = 0 to 180)",
  xlab="AgChem Formula",
  cex.lab=1.15, cex.axis=1.15, las=1, border="blue",
  col="red")
box()


par(ask=TRUE)
boxplot(AgChem.table$Larvae ~ AgChem.table$TimeOfDay.recode,
  horizontal=TRUE, # Horizontal boxes in this case.
  main="Horizontal Boxplot of Larvae Counts
  by Time-Of-Day",
  ylim=c(100,180),
  xlab="Larvae Counts (Limit = 0 to 180)",
  ylab="Time-Of-Day",
  cex.lab=1.15, cex.axis=1.15, las=3, border="blue",
  col="red")
box()


par(ask=TRUE)
boxplot(AgChem.table$Larvae ~ AgChem.table$TimeOfDay.recode,
  horizontal=FALSE,# Vertical boxes in this case.
  main="Vertical Boxplot of Larvae Counts
  by Time-Of-Day",
  ylim=c(120,160), # Note the values in this argument.
  ylab="Larvae Counts (Limit = 0 to 180)",
  xlab="Time-Of-Day",
  cex.lab=1.15, cex.axis=1.15, las=1, border="blue",
  col="red")
box()
```

The las argument was adjusted to force the axis label into a horizontal alignment, in relation to the axis. Review the output from help(par) to see the many options available for improvements to presentation.

Settings for cex.axis allow an easy-to-read presentation of the text. Verbose variable labels may be descriptive, but their placement on a graphic should be considered as they are worded. If all breakout group names do not appear in the graphic, adjust the cex.axis setting to a lower value. Or, consider using some type of abbreviation to shorten the label. As is nearly always the case with R, settings are generally a matter of personal preferences. Presentation should be as robust as possible, to take into account how the graphic shows at the back of a large lecture hall when projected to a screen with the distraction of filtered light coming in through window shades.

4.3.12   Vertical Violin Plots of the Summary Object Variable by Breakout Object Variables

As a new feature to these lessons, note how the two violin plot graphical images in this section have been saved as separate files. The violin plot of larvae counts by AgChem formula was saved in .jpeg graphical format. The violin plot of larvae counts by Time-of-Day was saved in .png graphical format.

Experiment with the different .bmp, .jpeg, .png, and .tiff graphics devices supported by R. Although .png format images seem to be frequently used, these many different formats all have slightly different features and the selection of a graphical format is often a matter of personal choice and specific needs.

It is certainly common to merely copy and paste a graphical image to a separate file, typically a file used for word processing. All features in the graphical image are usually retained. Even so, it is not always necessary to use copy and paste to take graphics generated in R and place them in a word processed document, presentation program, or on a Web page. This automated process to save graphical images is an option that should be considered, but R supports a wide range of options for nearly every action.

```
install.packages("UsingR")
library(UsingR)           # Load the UsingR package.
help(package=UsingR)      # Show the information page.
sessionInfo()             # Confirm all attached packages.

par(ask=TRUE)
savelab.cex  <- par(cex.lab=1.15)
saveaxis.cex <- par(cex.axis=1.15)
UsingR::simple.violinplot(AgChem.table$Larvae ~
                AgChem.table$Formula.recode,
   lty=1, lwd=6, col="cyan", ylim=c(0,180),
   ylab="Larvae Counts (Limit = 0 to 180)",
```

```
 xlab="AgChem Formula")
 title("Violin Plot of Larvae Counts
 by AgChem Formula")
par(savelab.cex) # Return to original setting.
par(saveaxis.cex)# Return to original setting.

par(ask=TRUE)
savelab.cex  <- par(cex.lab=1.15)
saveaxis.cex <- par(cex.axis=1.15)
UsingR::simple.violinplot(AgChem.table$Larvae ~
                  AgChem.table$Formula.recode,
 lty=1, lwd=6, col="cyan", ylim=c(0,180),
 ylab="Larvae Counts (Limit = 0 to 180)",
 xlab="AgChem Formula")
 title("Violin Plot of Larvae Counts
 by AgChem Formula")
dev.copy(jpeg,
   filename="JPEG_Larvae_by_Formula_Violin-Plot.jpeg",
   height=600,
   width=800,
   pointsize=12,
   bg="white")
dev.off()
readline("Press <Enter> to continue")
windows()
par(savelab.cex) # Return to original setting.
par(saveaxis.cex)# Return to original setting.

par(ask=TRUE)
savelab.cex  <- par(cex.lab=1.15)
saveaxis.cex <- par(cex.axis=1.15)
UsingR::simple.violinplot(AgChem.table$Larvae ~
                  AgChem.table$TimeOfDay.recode,
 lty=1, lwd=6, col="cyan", ylim=c(0,180),
 ylab="Larvae Counts (Limit = 0 to 180)",
 xlab="Time-Of-Day")
 title("Violin Plot of Larvae Counts by Time-Of-Day")
par(savelab.cex) # Return to original setting.
par(saveaxis.cex)# Return to original setting.

par(ask=TRUE)
savelab.cex  <- par(cex.lab=1.15)
saveaxis.cex <- par(cex.axis=1.15)
UsingR::simple.violinplot(AgChem.table$Larvae ~
                  AgChem.table$TimeOfDay.recode,
 lty=1, lwd=6, col="cyan", ylim=c(0,180),
```

```
    ylab="Larvae Counts (Limit = 0 to 180)",
    xlab="Time-Of-Day")
    title("Violin Plot of Larvae Readings by Time-Of-Day")
dev.copy(png,
    filename="PNG_Larvae_by_TimeOfDay_Violin-Plot.png",
    height=600,
    width=800,
    pointsize=12,
    bg="white")
dev.off()
readline("Press <Enter> to continue")
windows()
par(savelab.cex) # Return to original setting.
par(saveaxis.cex)# Return to original setting.
```

4.3.13 Beanplots of the Summary Object Variable by Breakout Object Variables

The next beanplot figure uses economy of presentation and puts all relevant beanplots into one comprehensive figure. This type of summary can be quite useful to help with a better understanding of how the data relate to each other, overall, and by groups. Although the beanplot is not yet a well-known tool, the type of figure generated below demands more use, for formative diagnostic purposes and for publication and presentation.

```
par(ask=TRUE)
beanplot::beanplot(list(all=AgChem.table$Larvae),
    Larvae ~ Formula.recode,
    AgChem.table,
    Larvae ~ TimeOfDay.recode,
    main="Vertical Beanplot of Larvae Counts
    by Breakout Groups, Beans do not Show",
    xlab="All Data and Then Breakout Groups",
    ylab="Larvae Counts",
    ll=0,  # Hide the beans.
    cex.lab=1.25,
    cex.axis=1.10,
    col.axis="darkblue",
    ylim=c(120,160),
    method="stack",
    overallline="mean",
    horizontal=FALSE,
```

```
col="red",
border = "blue",
show.names=TRUE,
log="")
```

4.3.14   Representation of Group Means and Confidence Intervals

Among the many uses of the gplots package, it is often used to display confidence intervals.

As a new feature to this lesson, look at the way the with() function and the subset() function have been used to gain descriptive statistics (Mean and SD) for specific breakout groups of the measured (e.g., counted) object variable, or AgChem.table$Larvae in this sample lesson. These descriptive statistics will be presented in the legend, to gain full utility for presentation.

```
install.packages("gplots")
library(gplots)          # Load the gplots package.
help(package=gplots)     # Show the information page.
sessionInfo()            # Confirm all attached
                         #   packages.

par(ask=TRUE)
gplots::plotmeans(Larvae ~ Formula.recode,
  main="Simple Representation of Larvae Counts
  (Mean and CI) by AgChem Formula")
```

Use == and not = to state the concept equals instead of assignment.

```
with(subset(AgChem.table, Formula.recode ==
  'AgChem Formula 1'),
  mean(Larvae, na.rm=TRUE))
with(subset(AgChem.table, Formula.recode ==
  'AgChem Formula 1'),
  sd(Larvae, na.rm=TRUE))

with(subset(AgChem.table, Formula.recode ==
  'AgChem Formula 2'),
  mean(Larvae, na.rm=TRUE))
with(subset(AgChem.table, Formula.recode ==
  'AgChem Formula 2'),
  sd(Larvae, na.rm=TRUE))

with(subset(AgChem.table, Formula.recode ==
  'AgChem Formula 3'),
  mean(Larvae, na.rm=TRUE))
```

```r
with(subset(AgChem.table, Formula.recode ==
  'AgChem Formula 3'),
  sd(Larvae, na.rm=TRUE))

savefont <- par(font=2) # Bold
par(ask=TRUE)
gplots::plotmeans(Larvae ~ Formula.recode,
  data=AgChem.table,
  main="Larvae Counts (Mean and CI) by AgChem Formula",
  font.main=1.25,
  bars=TRUE,
  p=0.95,
  xlab="AgChem Formula",
  ylab="Larvae Count (Limit = 0 to 180)",
  ylim=c(133,148),   # Adjust to mean values.
  mean.labels=FALSE, ci.label=TRUE, n.label=TRUE,
  pch=7, digits=1, col="darkred", barwidth=2,
  barcol="darkblue", font.lab=2, col.lab="darkgreen",
  cex.lab=1.25, font.axis=2, col.axis="darkgreen",
  connect=FALSE, use.t=TRUE)
legend("topright", # Legend placed in topright.
  legend = c(
  "145.44    Mean  AgChem Formula 1  ",
  "138.02    Mean  AgChem Formula 2  ",
  "135.34    Mean  AgChem Formula 3  ",
  "  8.61    SD    AgChem Formula 1  ",
  "  2.56    SD    AgChem Formula 2  ",
  "  3.43    SD    AgChem Formula 3  "),
    ncol=2,
    locator(1),
    xjust=1,
    text.col="black",
    inset=0.05,
    bty="n")
par(savefont) # Go back to default font settings.

par(ask=TRUE)
gplots::plotmeans(Larvae ~ TimeOfDay.recode,
  main="Simple Representation of Larvae Counts
  (Mean and CI) by Time-of-Day")

with(subset(AgChem.table, TimeOfDay.recode ==
  'AM Application'),
  mean(Larvae, na.rm=TRUE))
```

```
with(subset(AgChem.table, TimeOfDay.recode ==
  'AM Application'),
  sd(Larvae, na.rm=TRUE))

with(subset(AgChem.table, TimeOfDay.recode ==
  'PM Application'),
  mean(Larvae, na.rm=TRUE))
with(subset(AgChem.table, TimeOfDay.recode ==
  'PM Application'),
  sd(Larvae, na.rm=TRUE))

savefont <- par(font=2) # Bold
par(ask=TRUE)
gplots::plotmeans(Larvae ~ TimeOfDay.recode,
  data=AgChem.table,
  main="Larvae Counts (Mean and CI) by Time-Of-Day",
  font.main=1.25,
  bars=TRUE,
  p=0.95,
  xlab="Time-Of-Day",
  ylab="Larvae Counts (Limit = 0 to 180)",
  ylim=c(135,145),   # Adjust to mean values.
  mean.labels=FALSE, ci.label=TRUE, n.label=TRUE,
  pch=7, digits=1, col="darkred", barwidth=2,
  barcol="darkblue", font.lab=2, col.lab="darkgreen",
  cex.lab=1.25, font.axis=2, col.axis="darkgreen",
  connect=FALSE, use.t=TRUE)
legend("topleft", # Legend placed in topleft.
  legend = c(
  "137.14   Mean  AM Application   ",
  "142.06   Mean  PM Application   ",
  "  2.78   SD    AM Application   ",
  "  8.85   SD    PM Application   "),
    ncol=2,
    locator(1),
    xjust=1,
    text.col="black",
    inset=0.05,
    bty="n")
par(savefont) # Go back to default font settings.
```

The *Y*-axis was adjusted to `ylim=c(135,145)` instead of `ylim=c(0,180)` to allow sufficient room for the legend. Experiment with axis limits as figures are prepared, to improve presentation and ultimately understanding of outcomes by the typical viewer. R supports flexibility.

Finally, look at the way the gplots::plotmeans() function is used to make an even more complicated figure, that provides a view of Larvae counts for all six plots in this sample.

```
savelas  <- par(las=2)          # Perpendicular axis labels
savemai  <- par(mai=c(3.02,   # Bottom margin in inches
                      0.82,   # Left margin in inches
                      0.82,   # Top margin in inches
                      0.42)) # Right margin in inches
gplots::plotmeans(Larvae ~ interaction(
  Formula.recode,
  TimeOfDay.recode,
  sep ="    "),
  connect=TRUE,
  barwidth=2,
  barcol="darkblue",
  font.lab=2,
  col.lab="darkgreen",
  cex.lab=1.25,
  font.main=1.25,
  font.axis=2,
  col.axis="darkgreen",
  data=AgChem.table,
  xlab=" ",
  ylab="Larvae Counts",
  main="Larvae Counts:  AgChem Formula
  and Time-of-Day Breakouts")
par(savelas)
par(savemai)
```

4.3.15 Plot Breakout Object Values on a Continuum of the Summary Object Variable

Create a formula object of the various breakout values associated with the current analysis.

```
Breakouts.of.Formula <- (AgChem.table$Larvae ~
                          AgChem.table$Formula.recode)
Breakouts.of.Formula
class(Breakouts.of.Formula)
# Notice how Breakouts.of.Formula is a formula.

par(ask=TRUE)
plot.design(Breakouts.of.Formula,
  main="Larvae Counts by AgChem Formula (Mean)",
  fun=mean,          # Use the mean.
```

```
  xtick=TRUE,
  col="darkred",
  ylab="Mean",
  font.lab=2,
  cex.axis=1.15,
  lwd=3,
  cex.lab=1.15)

par(ask=TRUE)
plot.design(Breakouts.of.Formula,
  main="Larvae Counts by AgChem Formula (Median)",
  fun=median,        # Use the median.
  xtick=TRUE,
  col="darkred",
  ylab="Median",
  cex.axis=1.15,
  lwd=3,
  cex.lab=1.15)

Breakouts.of.TimeOfDay <- (AgChem.table$Larvae ~
                           AgChem.table$TimeOfDay.recode)
Breakouts.of.TimeOfDay
class(Breakouts.of.TimeOfDay)
# Notice how Breakouts.of.TimeOfDay is a formula.

par(ask=TRUE)
plot.design(Breakouts.of.TimeOfDay,
  main="Larvae Counts by Time-of-Day (Mean)",
  fun=mean,          # Use the mean.
  xtick=TRUE,
  col="darkred",
  ylab="Mean",
  cex.axis=1.15,
  lwd=3,
  cex.lab=1.15)

par(ask=TRUE)
plot.design(Breakouts.of.TimeOfDay,
  main="Larvae Counts by Time-of-Day (Median)",
  fun=median,        # Use the median.
  xtick=TRUE,
  col="darkred",
  ylab="Median",
  cex.axis=1.15,
  lwd=3,
  cex.lab=1.15)
```

Immediately below, note how an interaction algorithm (using mean in the first figure and median in the second figure) has been placed directly into the plot.design() function. Give attention to the ordering scheme used for the object variables.

```
par(ask=TRUE)
plot.design(Larvae ~ Formula.recode + TimeOfDay.
recode,
  data=AgChem.table,
  main="Mean Larvae Counts by AgChem Formula (Formula.recode)
  and also by Time-of-Day (TimeOfDay.recode)",
  fun=mean,          # Use the mean.
  xtick=TRUE,
  col="darkred",
  ylab="Mean",
  cex.axis=1.15,
  cex.lab=1.15)


par(ask=TRUE)
plot.design(Larvae ~ Formula.recode + TimeOfDay.recode,
  data=AgChem.table,
  main="Median Larvae Counts by AgChem Formula (Formula.recode)
  and also by Time-of-Day (TimeOfDay.recode)",
  fun=median,        # Use the median.
  xtick=TRUE,
  col="darkred",
  ylab="Median",
  cex.axis=1.15,
  cex.lab=1.15)
```

In the next set of figures, the interaction.plot() function is used to add to understanding of the data, specifically by breakout groups by mean values and then by median values.

```
savelwd        <- par(lwd=3)         # Heavy line
savefont       <- par(font=2)        # Bold
savecex.lab    <- par(cex.lab=1.25)  # Label
savecex.axis   <- par(cex.axis=1.25) # Axis
par(ask=TRUE)
interaction.plot(AgChem.table$Formula.recode,
                 AgChem.table$TimeOfDay.recode,
                 AgChem.table$Larvae,
  main="Interaction Plot of Mean Larvae Counts,
  AgChem Formula, and Time-of-Day",
  fun=mean, # Use mean instead of median.
  legend=TRUE,
  trace.label="Time-of-Day",
  fixed=TRUE,
```

```
  col=2:7,     # Multiple colors give contrast.
  lwd=4,
  xlab=" ",    # Blank label to allow for output.
  ylab="Mean Larvae Count",
  font.lab=2,
  ylim=c(120,160),
  xtick=TRUE)
par(savelwd)    # Return to original setting.
par(savefont)   # Return to original setting.
par(savecex.lab)  # Return to original setting.
par(savecex.axis) # Return to original setting.

savelwd        <- par(lwd=3)        # Heavy line
savefont       <- par(font=2)       # Bold
savecex.lab    <- par(cex.lab=1.25) # Label
savecex.axis   <- par(cex.axis=1.25)# Axis
par(ask=TRUE)
interaction.plot(AgChem.table$Formula.recode,
                 AgChem.table$TimeOfDay.recode,
                 AgChem.table$Larvae,
  main="Interaction Plot of Median Larvae Counts,
  AgChem Formula, and Time-of-Day",
  fun=median, # Use median instead of mean.
  legend=TRUE,
  trace.label="Time-of-Day",
  fixed=TRUE,
  col=2:7,     # Multiple colors give contrast.
  lwd=4,
  xlab=" ",    # Blank label to allow for output.
  ylab="Median Larvae Count",
  font.lab=2,
  ylim=c(120,160),
  xtick=TRUE)
par(savelwd)      # Return to original setting.
par(savefont)     # Return to original setting.
par(savecex.lab)  # Return to original setting.
par(savecex.axis) # Return to original setting.
  # There are no missing data.  If so, look at
  # help(interaction.plot) to see the impact of
  # missing data on this graphical function.
```

These two graphical images were only marginally embellished. The purpose of this display was to instead show the possible interaction effect, which may be in play if lines intersect in this type of graphic.

The syntax for production of these many images is easily modified to serve other comparisons. Once R syntax has been put into desired form, it is common practice to use and reuse this tested syntax for other applications.

If there were a desire to embellish the Interaction Plot of AgChem Formula by Time-of-Day, among many possible approaches, the subset() function could be used to obtain descriptive statistics for all six breakout groups of interest to this sample: (1AM) AgChem Formula 1 by AM Application, (1PM) AgChem Formula 1 by PM Application, (2AM) AgChem Formula 2 by AM Application, (2PM) AgChem Formula 2 by PM Application, (3AM) AgChem Formula 3 by AM Application, and (3PM) AgChem Formula 3 by PM Application. The Mean and SD could then be added to the Interaction Plot, as a legend, to provide even more useful information. This type of detail will improve presentation and guide the reader to see if there is any concern about interaction between the two object variables, AgChem Formula and Time-of-Day.

There are many ways to achieve this aim, using R. For this sample, the subset() function and Boolean declarations (e.g., Formula == 1 & TimeOfDay == 1) will be used to prepare breakouts of the desired groups. The pastecs::stat.desc() function will then be used to obtain a summary of relevant descriptive statistics for each of the six AgChem Formula by Time-of-Day breakout groups.

```
AgChem1AM.Larvae <- subset(AgChem.table$Larvae,
   Formula == 1 & TimeOfDay == 1)
AgChem1AM.Larvae

AgChem1PM.Larvae <- subset(AgChem.table$Larvae,
   Formula == 1 & TimeOfDay == 2)
AgChem1PM.Larvae

AgChem2AM.Larvae <- subset(AgChem.table$Larvae,
   Formula == 2 & TimeOfDay == 1)
AgChem2AM.Larvae

AgChem2PM.Larvae <- subset(AgChem.table$Larvae,
   Formula == 2 & TimeOfDay == 2)
AgChem2PM.Larvae

AgChem3AM.Larvae <- subset(AgChem.table$Larvae,
   Formula == 3 & TimeOfDay == 1)
AgChem3AM.Larvae

AgChem3PM.Larvae <- subset(AgChem.table$Larvae,
   Formula == 3 & TimeOfDay == 2)
AgChem3PM.Larvae

install.packages("pastecs")
library(pastecs)          # Load the pastecs package.
help(package=pastecs)     # Show the information page.
```

```
sessionInfo()              # Confirm all attached
                           packages.


pastecs::stat.desc(AgChem.table$Larvae)     # Mean, SD, etc.
```

With the pastecs::stat.desc() function, use the basic=FALSE
argument to obtain a list of the most pertinent descriptive statistics.

```
pastecs::stat.desc(AgChem1AM.Larvae, basic=FALSE)
pastecs::stat.desc(AgChem1PM.Larvae, basic=FALSE)
pastecs::stat.desc(AgChem2AM.Larvae, basic=FALSE)
pastecs::stat.desc(AgChem2PM.Larvae, basic=FALSE)
pastecs::stat.desc(AgChem3AM.Larvae, basic=FALSE)
pastecs::stat.desc(AgChem3PM.Larvae, basic=FALSE)
```

Given how the descriptive statistics for each of the six AgChem Formula by
Application Time-of-Day combinations have been calculated, use this information
to: (1) put the mean values into a new object (AgChem Formula Time, immediately below) and (2) make a simple figure of mean values for Larvae counts, using
the barplot() function. This action is another attempt to assist later comparisons
at a simple two-dimensional level of comparison.

```
AgChemFormulaTime <- c(137.20, 153.68, 137.26, 138.78,
   136.96, 133.72)
savelwd       <- par(lwd=3)        # Heavy line
savefont      <- par(font=2)       # Bold
savecex.lab   <- par(cex.lab=1.25) # Label
savecex.axis  <- par(cex.axis=1.25)# Axis
par(ask=TRUE)
barplot(AgChemFormulaTime,
   table(AgChemFormulaTime),
   main="AgChem Formula by Application Time-of-Day:
   Barplots of Mean Larvae Counts ",
   cex.main=1.25,
   horiz=FALSE,
   names.arg=c("AgChem1AM", "AgChem1PM",
               "AgChem2AM", "AgChem2PM",
               "AgChem3AM", "AgChem3PM"),
   xlab="AgChem Formula by Application Time-of-Day",
   ylab="Mean Larvae Counts",
   ylim=c(0,200),
   col="red")
savefamily   <- par(family="mono")
legend("top",
   legend = c(
   "AgChem1AM Mean = 137.20", "AgChem1PM Mean = 153.68",
```

```
    "AgChem2AM Mean = 137.26", "AgChem2PM Mean = 138.78",
    "AgChem3AM Mean = 136.96", "AgChem3PM Mean = 133.72"),
     ncol=3,
     locator(1),
     xjust=1,
     text.col="black",
     cex=1.15,
     inset=0.025,
     bty="n")
par(savelwd)
par(savefont)
par(savecex.lab)
par(savecex.axis)
par(savefamily)
```

Although the information in the barplot is useful, it would be better to additionally use the descriptive statistics for all six AgChem Formula by Application Time-of-Day breakout groups to generate a fully embellished interaction plot of AgChem Formula by Application Time-of-Day. The type of detail in this graphical figure should be quite telling, especially if there is an intersection between the two lines, indicating further interest in the possibility of some type of interaction between the two factor object variables, AgChem Formula and Application Time-of-Day.

A few features in this fully involved interaction plot need special attention here, in an attempt to prepare a figure that may be of publishable quality: (1) The Sys.time() function is added to the graphic, to provide a timestamp of date and time. This information may not be essential, but there are times when this would be useful given how multiple iterations of a figure or table are created, as data are updated. A timestamp provides useful context and tracking information. If needed, the timestamp can always be removed or hidden by using the R # comment character, as the last iteration is prepared. (2) Figures can easily have two or more legends, depending on need and preference. (3) Give special attention to the manual placement of the legend(s). It takes some trial-and-error to find the right coordinates to place a legend at a specific location. (4) Although it may seem a bit odd at first, look at how the getwd() function is added to the graphic, to provide a listing of the drive and directory for this graphic. This information may not be essential, but there are times when this would be useful given how multiple iterations of a figure or table are created, as data are updated. Knowing the drive and directory provides tracking information, which is certainly useful when a few weeks or more go by between use of the R syntax file and recall becomes fuzzy. If needed, the listing can always be removed as the last iteration is prepared. (5) Notice how ylim=c(115,160) was used instead of ylim=c(120,160). By adjusting the lower range on the scale, more room was allowed for the bottomright legend to show, without carryover into any other legend or output. As mentioned before, experiment and learn to adjust scales and range to accommodate final output.

```
png("PNG_Larvae_Interaction-Plot.png", width=700,
height=500)
savelwd        <- par(lwd=3)           # Heavy line
savefont       <- par(font=2)          # Bold
savecex.lab    <- par(cex.lab=1.25)    # Label
savecex.axis   <- par(cex.axis=1.25)   # Axis
par(ask=TRUE)
interaction.plot(AgChem.table$Formula.recode,
                 AgChem.table$TimeOfDay.recode,
                 AgChem.table$Larvae,
  main="Interaction Plot of Mean Larvae Counts,
  AgChem Formula, and Time-of-Day",
  fun=mean,
  legend=TRUE,
  trace.label="Time-of-Day",
  fixed=TRUE,
  col=2:7,      # Multiple colors give contrast.
  lwd=4,
  xlab=" ",     # Blank label to allow for output.
  ylab="Mean Larvae Count",
  font.lab=2,
  ylim=c(115,160),
  xtick=TRUE)
savefamily     <- par(family="mono") # Courier font
legend("bottomright",
  legend = c(
  "AgChem 1 AM Mean = 137.20 and SD = 2.66",
  "AgChem 1 PM Mean = 153.68 and SD = 2.05",
  "AgChem 2 AM Mean = 137.26 and SD = 2.30",
  "AgChem 2 PM Mean = 138.78 and SD = 2.60",
  "AgChem 3 AM Mean = 136.96 and SD = 3.32",
  "AgChem 3 PM Mean = 133.72 and SD = 2.73",
  "=======================================",
  "Overall     Mean = 139.60 and SD = 7.00"),
    ncol=1,
    locator(1),
    xjust=1,
    text.col="black",
    inset=0.05,
    bty="n")
par(savefamily)
legend(0.75, 119,        # Manual placement of the legend.
  legend = Sys.time(),   # Add date and time.
  cex=0.75,
  bty="n")
```

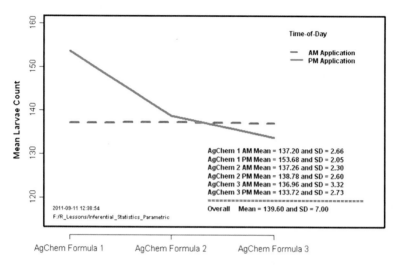

Fig. 4.1 Interaction plot of mean Larvae counts, AgChem Formula, and Time-Of-Day

```
legend(0.75,  117,        # Manual placement of the legend.
   legend = getwd(),      # Add directory path.
   cex=0.75,
   bty="n")
par(savelwd)       # Return to original setting.
par(savefont)      # Return to original setting.
par(savecex.lab)   # Return to original setting.
par(savecex.axis)  # Return to original setting.
dev.off()   # Here is another way to save graphical images.
```

Along with what was seen in the figure generated by using the barplot() function, this interaction.plot() figure provides some degree of indication that there are differences in larvae counts for the six breakout groups in this sample. Although statistics gained from the planned two-way ANOVA are needed to make exact declarations, at the least there seems to be evidence that the number of larvae associated with AgChem Formula 1 PM Application is widely different from the other breakout groups. Graphical presentation is essential to positive communication with the intended audience on outcomes. Reference to this interaction plot will be visited again, with more details added once the two-way ANOVA is completed (see Fig. 4.1).

4.4 Descriptive Analysis of the Data

Many different R functions that support descriptive statistics have been used to prepare the legends in the above figures. Even with these initial presentations, it is still necessary to carefully examine the dataset using summary and breakout descriptive statistics. Quality assurance demands a complete understanding of the data, through all stages of the data analysis process.

4.4.1 Summary Descriptive Statistics

For this presentation, a few (of the many) descriptive statistics will be prepared for the continuous object variable AgChem.table$Larvae, by itself and then by the factor object variables AgChem.table$Formula.recode and AgChem.table$TimeOfDay.recode. A variety of R functions that support descriptive statistics at the summary level include: mean(), sd(), median(), asbio::Mode(), summary(), pastecs::stat.desc(), psych::describe(), and the exceptionally useful epicalc::summ() function. The Hmisc package will then be used to provide even more options on the preparation of descriptive statistics. Each function has characteristics that may or may not be useful. Use all and then decide which functions meet individual requirements.

```
mean(AgChem.table$Larvae, na.rm=TRUE)
sd(AgChem.table$Larvae, na.rm=TRUE)
median(AgChem.table$Larvae, na.rm=TRUE)
length(na.omit(AgChem.table$Larvae)) # Note na.omit.
```

It may seem odd, at first, that R supports mean and median, but it does not support mode as another view of average. Key mode (AgChem.table$Larvae) and see what happens.

```
mode(AgChem.table$Larvae) # Key help(mode).
```

To resolve this problem, use the asbio::Mode() function to determine the mode (as a term for average, in this case the most frequently occurring value for AgChem.table$Larvae) of the object variable in question.

```
install.packages("asbio")
library(asbio)            # Load the UsingR package.
help(package=asbio)       # Show the information page.
sessionInfo()             # Confirm all attached packages.

asbio::Mode(AgChem.table$Larvae) # Note the use of capital M.
```

Given access to the asbio package, there are a few other trim-type functions here that may be useful, especially if there is any concern about assumptions associated with distribution of values representing Larvae counts.

```
mean(AgChem.table$Larvae)

asbio::trim.me(AgChem.table$Larvae, trim=0.025)
mean(asbio::trim.me(AgChem.table$Larvae, trim=0.025))
```

```
mean(asbio::trim.me(AgChem.table$Larvae, trim=0.050))
asbio::trim.me(AgChem.table$Larvae, trim=0.050)

asbio::skew(AgChem.table$Larvae)
asbio::kurt(AgChem.table$Larvae)

summary(AgChem.table$Larvae)
```

The pastecs package was previously installed and it will be used again to provide descriptive statistics.

```
pastecs::stat.desc(AgChem.table$Larvae,
  basic=TRUE, desc=TRUE, norm=TRUE, p=0.95)
```

```
# The psych package has many useful functions.
install.packages("psych")
library(psych)            # Load the psych package.
help(package=psych)       # Show the information page.
sessionInfo()             # Confirm all attached packages.

psych::describe(AgChem.table$Larvae)
```

The psych::describe() function is especially useful in that it can be applied against the entire dataset. The desired output can then be copied and pasted into a formal report. Merely delete the output applied against numbers that are instead factor-type object variables, which are marked with a * symbol. Even so, give attention to the values marked with a ⋆ symbol. They are useful as an additional tool used to check for possible errors.

```
psych::describe(AgChem.table)
```

The epicalc::summ() function provides output that is somewhat similar to psych::describe() when applied against the full set of data, organized in the data frame. As a simple quality assurance tool, be sure to always look at the min (minimum) and max (maximum) output, to look for data that may be out-of-range. Good research calls for constant checking and rechecking of the data. Never assume that one data check at the beginning of a project is sufficient. Quality assurance is a continuous process.

```
# The epicalc package was previously installed.
par(ask=TRUE)
epicalc::summ(AgChem.table$Larvae,
  by=NULL,
  graph=TRUE,
  box=TRUE,        # Generate a boxplot.
  pch=18,
  ylab="auto",
  main="Sorted Dotplot and Boxplot of AgChem.table$Larvae",
  cex.X.axis=1.15,
  cex.Y.axis=1.15,
```

```
font.lab=2,
dot.col="auto")
# Using the by=NULL argument, the summ() function does
# not provide breakout statistics, but instead provides
# descriptive statistics and an accompanying graphic
# for all values in AgChem.table$Larvae.

epicalc::summ(AgChem.table)
```

4.4.2 Breakout Descriptive Statistics

There are many ways to obtain breakout statistics. As previously shown, a series of brute force subset() function actions were used to identify group membership in defined subgroups, with functions for descriptive statistics applied against these subgroups. Also useful are the tapply() function, the psych::describe.by() function, and the epicalc::summ() function. Each selection has value and supports the ability to discern differences between breakout groups such as the three breakout groups for AgChem.table$Formula.recode associated with this sample lesson (AgChem Formula 1, AgChem Formula 2, and AgChem Formula 3) and the two application-specific breakout groups for AgChem.table$TimeOfDay.recode associated with this sample lesson (AM Application and PM Application).

Compare the output gained by use of the subset() function, the tapply() function, the psych::describe.by() function, and the epicalc::summ() function to see how these four functions can be used to generate breakout statistics. Then, determine which function has the best use.

```
# subset() Function
AgChem1.Larvae <- subset(AgChem.table$Larvae, Formula == 1)
AgChem1.Larvae

AgChem2.Larvae <- subset(AgChem.table$Larvae, Formula == 2)
AgChem2.Larvae

AgChem3.Larvae <- subset(AgChem.table$Larvae, Formula == 3)
AgChem3.Larvae

TimeOfDayAM.Larvae <- subset(AgChem.table$Larvae,
   TimeOfDay == 1)
TimeOfDayAM.Larvae

TimeOfDayPM.Larvae <- subset(AgChem.table$Larvae,
   TimeOfDay == 2)
TimeOfDayPM.Larvae
```

```
AgChem1AM.Larvae <- subset(AgChem.table$Larvae,
  Formula == 1 & TimeOfDay == 1)
AgChem1AM.Larvae

AgChem1PM.Larvae <- subset(AgChem.table$Larvae,
  Formula == 1 & TimeOfDay == 2)
AgChem1PM.Larvae

AgChem2AM.Larvae <- subset(AgChem.table$Larvae,
  Formula == 2 & TimeOfDay == 1)
AgChem2AM.Larvae

AgChem2PM.Larvae <- subset(AgChem.table$Larvae,
  Formula == 2 & TimeOfDay == 2)
AgChem2PM.Larvae

AgChem3AM.Larvae <- subset(AgChem.table$Larvae,
  Formula == 3 & TimeOfDay == 1)
AgChem3AM.Larvae

AgChem3PM.Larvae <- subset(AgChem.table$Larvae,
  Formula == 3 & TimeOfDay == 2)
AgChem3PM.Larvae

# pastecs::stat.desc() Function
pastecs::stat.desc(AgChem1.Larvae, basic=FALSE)
pastecs::stat.desc(AgChem2.Larvae, basic=FALSE)
pastecs::stat.desc(AgChem3.Larvae, basic=FALSE)

pastecs::stat.desc(TimeOfDayAM.Larvae, basic=FALSE)
pastecs::stat.desc(TimeOfDayPM.Larvae, basic=FALSE)

pastecs::stat.desc(AgChem1AM.Larvae, basic=FALSE)
pastecs::stat.desc(AgChem1PM.Larvae, basic=FALSE)
pastecs::stat.desc(AgChem2AM.Larvae, basic=FALSE)
pastecs::stat.desc(AgChem2PM.Larvae, basic=FALSE)
pastecs::stat.desc(AgChem3AM.Larvae, basic=FALSE)
pastecs::stat.desc(AgChem3PM.Larvae, basic=FALSE)
# tapply() Function
tapply(Larvae, Formula.recode,
  summary, digits=9, na.rm=TRUE, data=AgChem.table)

tapply(Larvae, TimeOfDay.recode,
  summary, digits=9, na.rm=TRUE, data=AgChem.table)

tapply(Larvae, list(Formula.recode, TimeOfDay.recode),
  mean,                 # Determine mean, only.
  digits=9, na.rm=TRUE, data=AgChem.table)
```

```
# psych::describe.by() Function
psych::describe.by(AgChem.table$Larvae,
                   AgChem.table$Formula.recode)

psych::describe.by(AgChem.table$Larvae,
                   AgChem.table$TimeOfDay.recode)

psych::describe.by(AgChem.table$Larvae,
  list(AgChem.table$Formula.recode,
       AgChem.table$TimeOfDay.recode))
  # There are two grouping variables.

psych::describe.by(AgChem.table$Larvae,
  list(AgChem.table$Formula.recode,
       AgChem.table$TimeOfDay.recode),
       mat=TRUE) # Output as a matrix.
  # There are two grouping variables.

par(ask=TRUE)
epicalc::summ(AgChem.table$Larvae,
   by=AgChem.table$Formula.recode,  # Note the use of by.
   graph=TRUE, # Use graph=TRUE,
   pch=20,     # if desired.
   ylab="auto",
   main="Sorted Dotplot of AgChem.table$Larvae
   by AgChem.table$Formula.recode",
   cex.X.axis=1.15, # Adjust presentation to best effect.
   cex.Y.axis=1.15, # Adjust presentation to best effect.
   font.lab=2,
   dot.col="auto")

par(ask=TRUE)
epicalc::summ(AgChem.table$Larvae,
   by=AgChem.table$TimeOfDay.recode,  # Note the use of by.
   graph=TRUE, # Use graph=TRUE,
   pch=20,     # if desired.
   ylab="auto",
   main="Sorted Dotplot of AgChem.table$Larvae
   by AgChem.table$TimeOfDay.recode",
   cex.X.axis=1.15, # Adjust presentation to best effect.
   cex.Y.axis=1.15, # Adjust presentation to best effect.
   font.lab=2,
   dot.col="auto")
```

The Hmisc package has many functions that support descriptive statistics. With planning, the Hmisc functions can be altered to allow for individual preferences in presentation. Of course, this is a major advantage of open source software such as R. A few Hmisc functions are shown below, but it is a valuable use of time to study documentation for the entire set of valuable functions supported by this frequently used R package.

```
install.packages("Hmisc")
library(Hmisc)              # Load the Hmisc package.
help(package=Hmisc)         # Show the information page.
sessionInfo()               # Confirm all attached packages.

# Hmisc Functions
Hmisc::describe(AgChem.table,
   digits=5,
   condense=FALSE,
   exclude.missing=FALSE,
   descript="Description of the data frame AgChem.table")
   # Note how NA (missing) is accommodated in this set of
   # analyses.

Hmisc::smean.sd(AgChem.table$Larvae,
   na.rm=TRUE)
   # Compute the mean and the standard deviation.
```

From among the many possible functions available for breakout descriptive statistics, the doBy::summaryBy() function is also useful, providing information in a fairly condensed format, similar to what might be expected from a crosstabs analysis.

```
install.packages("doBy")
library(doBy)               # Load the doBy package.
help(package=doBy)          # Show the information page.
sessionInfo()               # Confirm all attached packages.

# doBy::summaryBy() Function
doBy::summaryBy(Larvae ~
   Formula.recode +
   TimeOfDay.recode,
   data=AgChem.table,
      FUN=c(mean,sd),
      na.rm=TRUE,
      keep.names=TRUE,
      order=TRUE)

doBy::summaryBy(Larvae ~
   TimeOfDay.recode +
   Formula.recode,
   data=AgChem.table,
      FUN=c(mean,sd),
      na.rm=TRUE,
      keep.names=TRUE,
      order=TRUE)
```

It will be difficult to accommodate missing values for length. The enumerated function below takes care of this problem.

```
##################################################
descriptivefun <- function(x, ...){
  c(mean=mean(x, ...),          # Mean
    sd=sd(x, ...),              # Standard Deviation
    median=median(x, ...),      # Median
    length=length(x),           # Length
    min=min(x, ...),            # Minimum
    max=max(x, ...))            # Maximum
}
# This is a user-created (e.g., enumerated) function.
##################################################

doBy::summaryBy(Larvae ~
  Formula.recode +
  TimeOfDay.recode,
  data=AgChem.table,
    FUN=descriptivefun,
    na.rm=TRUE,
    keep.names=TRUE,
    full.dimension=FALSE,
    order=TRUE)
    # Use the enumerated
    # function.

doBy::summaryBy(Larvae ~
  TimeOfDay.recode +
  Formula.recode,
  data=AgChem.table,
    FUN=descriptivefun,
    na.rm=TRUE,
    keep.names=TRUE,
    full.dimension=FALSE,
    order=TRUE)
    # Use the enumerated
    # function.
```

The prettyR package should also be considered as a possible resource for descriptive statistics, especially the prettyR::brkdn() function.

```
install.packages("prettyR")
library(prettyR)          # Load the prettyR package.
help(package=prettyR)     # Show the information page.
sessionInfo()             # Confirm all attached packages.
```

```
# prettyR::brkdn() Function
prettyR::brkdn(Larvae ~ Formula.recode,
  data=AgChem.table,
  maxlevels=3,
  num.desc=c("mean", "sd", "valid.n"),
  width=25,
  round.n=2)

prettyR::brkdn(Larvae ~ TimeOfDay.recode,
  data=AgChem.table,
  maxlevels=2,
  num.desc=c("mean", "sd", "valid.n"),
  width=25,
  round.n=2)

prettyR::brkdn(Larvae ~ (Formula.recode +
                         TimeOfDay.recode),
  data=AgChem.table,
  maxlevels=6,
  num.desc=c("mean","sd","valid.n"),
  width=25,
  round.n=2)
  # Observe how the summary statistic
  # is provided, and then breakouts.

prettyR::brkdn(Larvae ~ (TimeOfDay.recode +
                         Formula.recode),
  data=AgChem.table,
  maxlevels=6,
  num.desc=c("mean","sd","valid.n"),
  width=25,
  round.n=2)
  # Observe how the summary statistic
  # is provided, and then breakouts.
```

In what must seem to be a growing list of functions, the `plotrix::brkdn.plot()` deserves attention for detail and presentation.

```
install.packages("plotrix")
library(plotrix)          # Load the plotrix package.
help(package=plotrix)     # Show the information page.
sessionInfo()             # Confirm all attached packages.

# plotrix::brkdn.plot() Function
savelwd        <- par(lwd=3)         # Heavy line
savefont       <- par(font=2)        # Bold
```

```
savecex.lab    <- par(cex.lab=1.25) # Label
savecex.axis   <- par(cex.axis=1.25)# Axis
savetck        <- par(tck=0.05)     # Tick Marks
par(ask=TRUE)
plotrix::brkdn.plot(
  "Larvae", "TimeOfDay.recode", "Formula.recode",
  AgChem.table,
  mct="mean",
  dispbar=TRUE,
  main="Breakdown Plot of Larvae Counts by Factors
  Formula.recode and TimeOfDay.recode (Mean):
  plotrix::brkdn.plot()",
  xlab="Formula.recode",
  ylab="Larvae Reading",
  type="b",
  pch=1:4,
  lty=1:4,
  col=c("blue","red","green", "black"),
  staxx=TRUE)
par(savelwd)          # Go back to default settings.
par(savefont)         # Go back to default settings.
par(savecex.lab)      # Go back to default settings.
par(savecex.axis)     # Go back to default settings.
par(savetck)          # Go back to default settings.
colorfill <- c("blue","red","green","black")
legend(locator(1),
  levels(AgChem.table$TimeOfDay.recode),
  fill=colorfill)
  # Give attention to the color ordering.
  # Never assume that the ordering reflects
  # the values but instead confirm that the
  # colors reflect true values.
savefamily <- par(family = "mono")    # Courier-type
                                        font.
savefont   <- par(font=2)             # Bold
legend("topright",
  legend = c(
    " Larvae Count by Agricultural Chemical Formula ",
    " and by Application Time-of-Day                 ",
    " ===============================================",
    " AgChem    Time-of-                             ",
    " Formula   Day       Mean    SD      N   Min  Max",
    " -----------------------------------------------",
    "                                                ",
    "    1       AM        137.20  2.6573  50  130  140",
```

```
          "    1    PM       153.68   2.0548   50   147   156",
          "    2    AM       137.26   2.3018   50   131   140",
          "    2    PM       138.78   2.5974   50   134   149",
          "    3    AM       136.96   3.3195   50   122   140",
          "    3    PM       133.72   2.7258   50   127   137",
          " ----------------------------------------------------"),
     ncol=1,
     locator(1),
     xjust=1,
     text.col="black",
     cex=0.80,
     inset=0.01,
     bty="n")
par(savefamily)
par(savefont)
```

As a reminder, a color-coded legend is of limited value to anyone who has trouble distinguishing one color from another. Plus, the details of the text-based descriptive legend provide even greater detail than a color-coded legend can provide. Use caution in the preparation of a figure, however. A figure cannot be so busy that it is difficult to read and understand.

4.4.3 Contingency Tables

Along with descriptive statistics and the production of graphical figures, it is also desirable to have a table or some type of summary presentation, to organize output. The table() and ftable() functions will serve quite nicely for the production of a simple table of frequency distributions.

```
table(AgChem.table$Formula.recode,      # Row
      AgChem.table$TimeOfDay.recode)     # Column

table(AgChem.table$TimeOfDay.recode,     # Row
      AgChem.table$Formula.recode)       # Column

ftable(AgChem.table$Formula.recode,      # Row
       AgChem.table$TimeOfDay.recode)    # Column

ftable(AgChem.table$TimeOfDay.recode,    # Row
       AgChem.table$Formula.recode)      # Column
```

There are times, however, when a far more detailed table, typically referred to as a contingency table, is desired. To achieve this aim, the gmodels::CrossTable() function is a good first choice. It can be simple in output or detailed, all by adjusting the selected arguments.

```
install.packages("gmodels")
library(gmodels)          # Load the gmodels package.
help(package=gmodels)     # Show the information page.
sessionInfo()             # Confirm all attached
                            packages.

gmodels::CrossTable(AgChem.table$Formula.recode,     # Row
                    AgChem.table$TimeOfDay.recode,   # Column
    digits=2,
    max.width=5,
    expected=FALSE,        # Note the use of FALSE.
    prop.r=FALSE,          # Note the use of FALSE.
    prop.c=FALSE,          # Note the use of FALSE.
    prop.t=FALSE,          # Note the use of FALSE.
    prop.chisq=FALSE,      # Note the use of FALSE.
    chisq=FALSE,           # Note the use of FALSE.
    missing.include=TRUE,  # Note the use of TRUE.
    format="SPSS")
    # Simple output.

gmodels::CrossTable(AgChem.table$Formula.recode,     # Row
                    AgChem.table$TimeOfDay.recode,   # Column
    digits=2,
    max.width=5,
    expected=TRUE,         # Note the use of TRUE.
    prop.r=TRUE,           # Note the use of TRUE.
    prop.c=TRUE,           # Note the use of TRUE.
    prop.t=TRUE,           # Note the use of TRUE.
    prop.chisq=TRUE,       # Note the use of TRUE.
    chisq=TRUE,            # Note the use of TRUE.
    missing.include=TRUE,  # Note the use of TRUE.
    format="SPSS")
    # Detailed output.
```

When preparing contingency tables, it is common to refer to organization in a Row by Column format. Unless there is a compelling reason to do otherwise, it is then common to make the factor object variable with the greatest number of factors as the factor that represents rows and to then make the factor object variable with the fewest number of factors as the factor that represents columns. For this sample, note how Formula.recode (three breakout levels) is presented as rows and TimeOfDay.recode (two breakout levels) is presented as columns. When the number of breakout groups is not the same for rows and columns, the above placement of rows and columns is done just to improve presentation on the screen or page, reducing the need to edit output or to put output into landscape mode for any formal report.

4.5 Use R for Two-Way Analysis of Variance (ANOVA)

The data have now been brought into this R session, data were organized and labeled to accommodate human cognition, graphical images have been produced at the summary level and breakout levels, descriptive statistics were generated at the summary level and breakout levels, and data have also been organized into simple and complex contingency tables. With all actions now in final form, the data are ready for inferential analyses, such as two-way ANOVA for this sample.

The task now is to use two-way ANOVA, as supported in R by the many available packages and functions, to determine if there are statistically significant differences in the number of Larvae between (1) the three AgChem Formula breakout groups, (2) the two Time-of-Day breakout groups, and (3) interaction between AgChem Formula breakout groups and Time-of-Day breakout groups. The descriptive statistics and graphical images serve as a basis for an initial suggestion that differences may or may not exist, but only an inferential test can provide the precise assessment needed for informed judgment.

R supports many possible ways to perform a two-way ANOVA. A few methods that support two-way ANOVA are detailed below.

4.5.1 Two-Way ANOVA: `aov()` Function

Use the formula for a two-way factorial design ANOVA, which is typically represented as:

```
fit1 <- aov(y ~ A + B + A:B, data=dataframe)
summary(fit1)

fit2 <- aov(y ~ A*B, data=dataframe)
summary(fit2)

y   = Measured datum, (e.g., weight, exam score, etc.)
A   = Factor Variable A (e.g., Gender, Race-Ethnicity, etc.)
B   = Factor Variable B (e.g., Soil Type, Breed Type, etc.)
A:B = Interaction of Factor Variable A and Factor Variable B
```

Both ANOVA formulas yield the same result, but the following formula (e.g., algorithm) is likely more illustrative:

```
fit1 <- aov(y ~ A + B + A:B, data=dataframe)

Sample3.fit1 <- aov(Larvae ~ Formula.recode +
                             TimeOfDay.recode +
                             Formula.recode:TimeOfDay.recode,
     data=AgChem.table)
summary(Sample3.fit1)
```

```
#                         Sum  Mean F
#                     Df  Sq   Sq   value           Pr(>F)
# Formula.recode       2  5475 2737 393   <0.0000000000000002 ***
# TimeOfDay.recode     1  1815 1815 261   <0.0000000000000002 ***
# Formula.recode:
#   TimeOfDay.recode   2  5294 2647 380   <0.0000000000000002 ***
# Residuals          294  2047    7
# ---
# >
```

```
Sample3.fit2 <- aov(Larvae ~ Formula.recode*TimeOfDay.recode,
  data=AgChem.table)
summary(Sample3.fit2)
```

```
#                         Sum  Mean F
#                     Df  Sq   Sq  ·value           Pr(>F)
# Formula.recode       2  5475 2737 393   <0.0000000000000002 ***
# TimeOfDay.recode     1  1815 1815 261   <0.0000000000000002 ***
# Formula.recode:
#   TimeOfDay.recode   2  5294 2647 380   <0.0000000000000002 ***
# Residuals          294  2047    7
# ---
# >
```

To gain a sense of the descriptive statistics (summary and breakout), use the model.tables() function for another view of Grand Mean, Mean, and N for each cell in the factorial table.

```
print(model.tables(Sample3.fit1,"means"), digits=3)
print(model.tables(Sample3.fit2,"means"), digits=3)
```

A review of this output brings immediate attention to the Larvae count for AgChem Formula 1 PM application (Mean = 153.7) and how it greatly exceeds the Mean for the five other breakout cells. Then, compare the two-way ANOVA output and the finding that $p <= 0.01$ for AgChem Formula, $p <= 0.01$ for Time-of-Day, and $p <= 0.01$ for interaction between AgChem Formula and Time-of-Day.

```
#     Formula.recode:TimeOfDay.recode
#                     TimeOfDay.recode

#     Formula.recode     AM Application PM Application
#        AgChem Formula 1 137.2          153.7
#        AgChem Formula 2 137.3          138.8
#        AgChem Formula 3 137.0          133.7
```

As a throwaway diagnostic, use the plot.design() function to see general trends for Larvae counts by each breakout group.

```
par(ask=TRUE)
plot.design(Larvae ~ Formula.recode + TimeOfDay.
  recode, data=AgChem.table)
```

Then, use the `interaction.plot()` function for more details, but now focusing on the object variables of greatest interest.

```
par(ask=TRUE)
interaction.plot(AgChem.table$Formula.recode,
                 AgChem.table$TimeOfDay.recode,
                 AgChem.table$Larvae,
  main="Interaction Plot:  AgChem Formula, Time-of-Day,
  and Larvae",
  font.lab=2, col=2:9, lty="solid", lwd=6)
```

Although the above aov() function meets the basic need for a two-way ANOVA and it would be possible to provide a brief summary of outcomes, there are many other R functions that can go well beyond this basic information. A few of these other R functions will be demonstrated below. These additional R functions provide a rich understanding of the data and they provide additional insight into interaction(s). These insights should not be overlooked and are always of interest to experienced researchers.

4.5.2 Additional R Packages that Support Two-Way ANOVA

```
install.packages("s20x")
library(s20x)             # Load the s20x package.
help(package=s20x)        # how the information page.
sessionInfo()             # Confirm all attached packages.

saveaxis <- par(cex.axis=0.95)
par(ask=TRUE)
s20x::boxqq(Larvae ~ Formula.recode,
  data=AgChem.table)
par(saveaxis)
  # Labels on the X axis were made smaller so that
  # all text shows in the figure.
saveaxis <- par(cex.axis=0.95)
par(ask=TRUE)
s20x::boxqq(Larvae ~ TimeOfDay.recode,
  data=AgChem.table)
par(saveaxis)
  # Labels on the X axis were made smaller so that
  # all text shows in the figure.
```

The `s20x::interactionPlots()` function does not work correctly if there are missing data. This issue is not a concern in this sample. If it were, the dataset would need to be adjusted so that each case has a full set of data, but this action is not always desirable.

```
par(ask=TRUE)
s20x::interactionPlots(Larvae ~ Formula.recode+
                                TimeOfDay.recode,
  AgChem.table,
  xlab="AgChem Formula",
  xlab2="Time-of-Day",
  ylab="Larvae Counts (Limit = 0 to 160)",
  type="hsd",
  tick.length=0.1,
  interval.distance=0.1,
  col.width=3/4,
  xlab.distance=1,
  xlen=1.25,
  ylen=1.25)
  # There are four options to the type= argument:
  # tukey, hsd, lsd, and ci.

s20x::crosstabs(~ Formula.recode + TimeOfDay.recode,
  AgChem.table)
```

Going back to the prior creation of Sample3.fit1 and Sample3.fit2, notice how the function s20x::summary2way() can be used to generate a two-way ANOVA table.

```
s20x::summary2way(Sample3.fit1,
  page = "table", digit = 5,
  conf.level = 0.95, print.out = TRUE)

s20x::summary2way(Sample3.fit1,
  page = "means", digit = 5,
  conf.level = 0.95, print.out = TRUE)



s20x::summary2way(Sample3.fit2,
  page = "table", digit = 5,
  conf.level = 0.95, print.out = TRUE)

s20x::summary2way(Sample3.fit2,
  page = "means", digit = 5,
  conf.level = 0.95, print.out = TRUE)
```

The page = "means" argument provides detailed summary statistics for overall, by the different factor object variables, and by combinations of the different factor object variables.

The output from use of the s20x package is especially rich and it should always be considered, both for the two-way ANOVA output as well as the way breakout descriptive statistics are presented.

```
install.packages("car")
library(car)              # Load the car package.
help(package=car)         # Show the information page.
sessionInfo()             # Confirm all attached packages.
```

Again, use the previously created objects Sample3.fit1 and Sample3.fit2 and use them again, now with the car package.

```
car::Anova(Sample3.fit1)      # Make a Two-Way ANOVA table.


car::Anova(Sample3.fit1, type="II")


car::Anova(Sample3.fit1, type="III")
  # Review car::Anova documentation to see possible
  # cautions with type="III" and why it should be used
  # sparingly and then only after consideration -- if
  # used at all.

car::Anova(Sample3.fit2)      # Make a Two-Way ANOVA table.


car::Anova(Sample3.fit2, type="II")


car::Anova(Sample3.fit2, type="III")
  # Review car::Anova documentation to see possible
  # cautions with type="III" and why it should be used
  # sparingly and then only after consideration -- if
  # used at all.
```

4.5.3   Outcome to Sample 3

The many R functions demonstrated in this sample have been used to explore the data. It is a minimal set of analyses for a two-way ANOVA that only provides descriptive statistics, a simple plot or two, and the corresponding ANOVA output table with F-values and p-values. Data are too valuable and outcomes are too important to conduct only the minimal actions.

It is important to examine the null hypothesis for this sample once again: there is no difference between AgChem formula, AgChem application time-of-day, and interaction between AgChem formula and AgChem application time-of-day regarding larvae counts for an unidentified insect at randomly selected areas in a field with otherwise uniform features ($p <= 0.05$). As prepared, this null hypothesis address three separate, but related, questions: (1) in regard to Larvae counts, are the population means for AgChem Formula (the first factor object variable in question: AgChem Formula 1, AgChem Formula 2, and AgChem Formula 3) equal? (2) In regard to Larvae counts, are the population means for Time-of-Day (the second factor object variable in question; AM Application and PM Application) equal? (3) In regard to Larvae counts, is there any observable interaction between AgChem Formula and Time-of-Day?

The two-way ANOVA of Larvae counts by AgChem Formula and by Time-of-Day follows, as generated by the prior s20x::summary2way() function and shown again, with some of the embellishments of the s20x:summary2way() function masked (sum square and mean square were deleted from this printout) to promote a more concise presentation of outcomes:

```
# > s20x::summary2way(Sample3.fit1)
# ANOVA Table:
#                               Df    F-statistic   p-value
# Formula.recode               2      393.1547      0
# TimeOfDay.recode             1      260.73779     0
# Formula.recode:TimeOfDay.
  recode                       2      380.1945      0
# Residuals                    294
# Total                        299
# >
```

By reviewing this output and using p <= 0.05 as the criterion *p*-value for the null hypothesis, consider the following summary.

4.5.3.1   AgChem Formula

There is a statistically significant difference (p <= 0.05) in Mean Larvae counts by the three AgChem Formula breakout groups (df = 2, $F = 393.15$, calculated *p* is listed as either 0 or 0.0000000000000002). The descriptive statistics that support this observation are from the s20x::summary1way() function, but these descriptive statistics have been prepared in this lesson by using many different functions:

```
# Numeric Summary:
#                    Sample size    Mean   Median   Std Dev
# All Data               300       139.60   138.0   6.9955
# AgChem Formula 1       100       145.44   143.5   8.6121
# AgChem Formula 2       100       138.02   138.0   2.5583
# AgChem Formula 3       100       135.34   136.0   3.4325
```

However, to be precise, it is best to now apply the post hoc TukeyHSD() function, which is associated with one-way ANOVA, against Larvae counts and AgChem Formula.

```
Larvae.by.AgChemFormula.OnewayANOVA <-
   aov(Larvae ~ Formula.recode,
   data=AgChem.table)

Larvae.by.AgChemFormula.OnewayANOVA

summary(Larvae.by.AgChemFormula.OnewayANOVA)

TukeyHSD(Larvae.by.AgChemFormula.OnewayANOVA)
```

```
# Multiple comparisons.
# This analysis accommodates missing data.
par(ask=TRUE)
plot(TukeyHSD(Larvae.by.AgChemFormula.OnewayANOVA),
   main="\n\nTukeyHSD Mean Comparison of AgChem Formula
   and Larvae Counts",
   cex.main=0.95,
   cex.lab=0.55,
   cex.axis=0.55,
   col.axis="darkblue",
   las=3,
   font.lab=2,
   font.axis=2,
   col="red")
# Plot a graphical reinforcement of Mean Comparison
# results, based on the TukeyHSD() function.
```

The output from use of the TukeyHSD() function provides further evidence needed to determine that there is a statistically significant difference (p <= 0.05) in Larvae counts by combinations of the three AgChem Formula breakout groups, as demonstrated in the p-values below.

```
# > TukeyHSD(Larvae.by.AgChemFormula.OnewayANOVA)
#    Tukey multiple comparisons of means
#      95% family-wise confidence level
#
# $Formula.recode
#                                          p adj
# AgChem Formula 2-AgChem Formula 1   0.00000
# AgChem Formula 3-AgChem Formula 1   0.00000
# AgChem Formula 3-AgChem Formula 2   0.00211
```

These three p-values, with each showing a p-value of less than 0.05, provide evidence that there is a statistically significant difference in Mean Larvae counts between:

```
Formula 1 (Mean = 145.44) and Formula 2 (Mean = 138.02)
Formula 1 (Mean = 145.44) and Formula 3 (Mean = 135.34)
Formula 2 (Mean = 138.02) and Formula 3 (Mean = 135.34)
```

4.5.3.2 Application Time-of-Day

There is a statistically significant difference (p <= 0.05) in Mean Larvae counts by the two Time-of-Day breakout groups (df = 1, $F = 260.74$, calculated p is listed as either 0 or 0.0000000000000002). The descriptive statistics that support this observation are from the s20x::summary1way() function, but these descriptive statistics have been prepared in this lesson by using many different functions:

```
# Numeric Summary:
#                      Sample size     Mean   Median   Std Dev
# All Data                     300   139.60      138    6.9955
# AM Application               150   137.14      138    2.7758
# PM Application               150   142.06      139    8.8494
```

However, to be precise, it is best to now apply the post hoc TukeyHSD() function, which is associated with one-way ANOVA, against Larvae counts and application Time-of-Day.

```
Larvae.by.TimeOfDay.OnewayANOVA <-
  aov(Larvae ~ TimeOfDay.recode,
  data=AgChem.table)


Larvae.by.TimeOfDay.OnewayANOVA


summary(Larvae.by.TimeOfDay.OnewayANOVA)


TukeyHSD(Larvae.by.TimeOfDay.OnewayANOVA)
  # Multiple comparisons.
  # This analysis accommodates missing data.

par(ask=TRUE)
plot(TukeyHSD(Larvae.by.TimeOfDay.OnewayANOVA),
  main="\n\nTukeyHSD Mean Comparison of Application
  Time-of-Day and Larvae Counts",
  cex.main=0.95,
  cex.lab=0.95,
  cex.axis=0.95,
  col.axis="darkblue",
  las=3,
  font.lab=2,
  font.axis=2,
  col="red")
  # Plot a graphical reinforcement of Mean Comparison
  # results, based on the TukeyHSD() function.
```

The output from use of the TukeyHSD() function provides the evidence needed to determine that there is a statistically significant difference ($p \leq 0.05$) in Larvae counts by the AM and PM application Time-of-Day scenario, as demonstrated in the p-values below.

```
# > TukeyHSD(Larvae.by.TimeOfDay.OnewayANOVA)
#    Tukey multiple comparisons of means
```

```
#       95% family-wise confidence level
#
# $TimeOfDay.recode
#                                        p adj
# PM Application-AM Application          0
#
```

This *p*-value, showing as a *p*-value of less than 0.05, provide evidence that there is a statistically significant difference in Mean Larvae counts between:

```
AM Application (Mean = 137.14) and
PM Application (Mean = 142.06)
```

4.5.3.3 Interaction Between AgChem Formula and Time-of-Day

In view of Larvae counts, the interaction effect between AgChem Formula and Time-of-Day was significant (calculated $p <= 0.0000000000000002$ which is less than the criterion $p <= 0.05$).

```
Sample3.fit1 <- aov(Larvae ~ Formula.recode +
                            TimeOfDay.recode +
                            Formula.recode:TimeOfDay.recode,
   data=AgChem.table)
summary(Sample3.fit1)


Sample3.fit2 <- aov(Larvae ~ Formula.recode*TimeOfDay.recode,
   data=AgChem.table)
summary(Sample3.fit2)

# > summary(Sample3.fit2)
#                                        Pr(>F)
# Formula.recode                  <0.0000000000000002 ***
# TimeOfDay.recode                <0.0000000000000002 ***
# Formula.recode:TimeOfDay.recode <0.0000000000000002 ***
# Residuals
# ---
```

Given this finding of significant interaction, it is useful to go back to the many descriptive statistics (summary and breakout) of Larvae Count (Mean, SD, and *N*) by AgChem Formula and by Time-of-Day:

```
s20x::summary2way(Sample3.fit1,
  page = "means", digit = 5,
  conf.level = 0.95, print.out = TRUE)

#                     TimeOfDay.recode
#                     AM          PM
# Formula.recode      Application  Application  Formula.recode
# AgChem Formula 1 137.20          153.68          145.44
```

```
#  AgChem Formula 2 137.26        138.78        138.02
#  AgChem Formula 3 136.96        133.72        135.34
#  TimeOfDay.recode 137.14        142.06        139.60
```

A careful examination of this table provides possible evidence of how to react to this finding of significant interaction. Specifically: There were statistically significant differences ($p <= 0.05$) in Mean Larva counts by AgChem Formula and also by application Time-of-Day.

Look carefully at the Mean Larvae Count for Formula 1 PM application (Mean = 153.68). Even a beginning researcher would consider the Mean number of Larvae for this breakout as a finding of importance. It is unknown to the researcher if the Larvae referenced in this lesson represent either desirable or undesirable insects, but whatever their status, the conditions in this plot (Formula 1 PM application) certainly favored this insect compared to conditions in the other plots as well as the overall grand Mean (Mean = 139.60) of Larvae counts.

To supplement this casual observation of cell means, look again at the figures that were previously generated, using the `interaction.plot()` function and the `plotrix::brkdn.plot()` function. Individually and collectively, these graphical figures are quite telling. Notice how the lines cross in these figures, giving some degree of indication about interaction.

Accordingly, with a fair degree of certainty, the sample provided evidence that, for this dataset, there are differences in Mean Larvae counts by AgChem Formula, by application Time-of-Day, and there is also interaction between AgChem Formula and application Time-of-Day. Production costs and especially rising global commodity prices such as $100 USD per barrel Brent crude oil and $8 USD per bushel No. 2 yellow corn have placed extreme pressure on profit margins in contemporary agricultural operations. Agricultural production demands the type of information presented in this lesson, often with the difference between profit and loss for those producers who use this type of information to enhance management. There may also be environmental considerations, given the nature of the different additives possibly showing in the three AgChem formulations.

These conclusions are stated with the constant caution that tight controls and replication are both essential parts of the overall research process. Consistent outcomes across various scenarios, various locations, soil types, weather conditions, application methods, and crops would be needed before any definitive conclusions could be offered and new management techniques put into place. This sample is merely one attempt in this overall assessment of factors impacting agricultural production.

Even so, this sample was a useful example of interaction between chemicals, which is certainly an issue in the biological sciences. Delivery methods (AgChem Formula) and environmental factors (application Time-of-Day) are only a few of the many issues of importance to contemporary food production, preparation, and transportation.

4.5.4   Consideration of the Data from a Nonparametric Perspective

The larvae counts for this sample lesson are accurate. IPM scouts have specialized training and adequate supervision to be sure that accepted protocols are followed and that the measures they generate are reliable and valid. Further, these counts are considered an acceptable proxy for weight, given the uniform patterns of timing and insect growth in this sample lesson.

Even with this assurance, there may be some who would suggest that the larvae counts should be viewed from a nonparametric perspective. Using this view, a larvae count of 155 would be considered more than a larvae count of 135, but the granularity of the counts would be suspect and the larvae counts would be put into rank order. Again, it is argued that this is not the case for this sample lesson, but it would still be a useful exercise to conduct a few inferential tests on the data, exploring the interplay of Larvae counts by AgChem Formula and by application Time-of-Day from this nonparametric perspective.

Although there are a few specialized R packages that could be used, the `asbio` package will be used for these initial additional tests.

```
install.packages("asbio")
library(asbio)            # Load the asbio package.
help(package=asbio)       # Show the information page.
sessionInfo()             # Confirm all attached packages.
```

Apply the Brunner–Dette–Munk (BDM) test as an alternate to use of the otherwise nonparametric Kruskal–Wallis one-way ANOVA.

```
asbio::BDM.test(
   Y=AgChem.table$Larvae,
   X=AgChem.table$Formula)

asbio::BDM.test(
   Y=AgChem.table$Larvae,
   X=AgChem.table$TimeOfDay)
```

Apply a pairwise post hoc ANOVA comparison test, similar to comparisons gained by using the `TukeyHSD()` function.

```
asbio::Pairw.test(
  x=AgChem.table$Formula,
  y=AgChem.table$Larvae,
  method="Tukey")
  # x and y must be lowercase.

asbio::Pairw.test(
  x=AgChem.table$TimeOfDay,
```

```
y=AgChem.table$Larvae,
method="Tukey")
# x and y must be lowercase.
```

Apply a multiple pairwise comparison test (review Kruskal–Wallis one-way ANOVA), similar to comparisons using the TukeyHSD() function.

```
asbio::KW.multi.comp(
  Y=AgChem.table$Larvae,
  X=AgChem.table$Formula,
  conf=.95)
```

```
asbio::KW.multi.comp(
  Y=AgChem.table$Larvae,
  X=AgChem.table$TimeOfDay,
  conf=.95)
```

Apply the Brunner–Dette–Munk (BDM) test as an alternate to use of the otherwise nonparametric Friedman two-way ANOVA.

```
asbio::BDM.2way(
  Y =AgChem.table$Larvae,
  X1=AgChem.table$Formula,
  X2=AgChem.table$TimeOfDay)
```

Use these tests against the data see if there is any meaningful difference in how outcomes are interpreted, whether the data are viewed from a parametric view or a nonparametric view.

4.5.5   Optional Housekeeping at End-of-Session

For simple analyses, the R session may have only a few objects available for use. In a lengthy R session, such as the above set of analyses, there are possibly many objects available for use. Now is the time to consider if some housekeeping chores are needed for this R session, using either the detach() function, the rm() function, the remove.packages() function, or some combination of all three functions:

Is it desirable to use the detach() function, to detach a data frame object that was previously attached? If the detach() function were used, the data frame object is still found in the active R session and it is still available for later use. Use of the detach() function does not remove an object, such as a data frame. Key help(detach) to learn more about this function.

Is it best to use the rm() function and instead remove some (or all) objects that are currently found in the R session? Remember that there is no warning when the rm() function is used, so think about this action before objects are removed. If the rm() function were used, is it desirable to first save the data and objects found in the current session? Key help(rm) to learn more about this function.

Is it desirable to use the remove.packages() function when the functions in the imported package are no longer needed? Some packages use function names that are also used in other packages, which creates possible conflicts. Or, some packages use at least similar function names, compared to function names in other packages. These similar names can be inadvertently used through a simple typo. Conflicts can be avoided by using a formal function call, such as package_name::function_name() but there are still good reasons why packages should be removed when are they are no longer needed. They can be easily brought back, when needed. Key help(remove.packages) to learn more about this function.

These (and other questions) are of no small importance since clutter can cause confusion and perhaps conflict with later analyses, especially if common names are used more than once to identify objects (e.g., SubjectID, SubjectName, Gender, Region, etc.), or functions from external packages are used without a formal function call (e.g., If summ() is used instead of epicalc::summ()). It is only too easy to make a typo and use sum() instead of summ() which of course is not the intent of the function.

For this lesson, the rm(list = ls()) function has been used to remove all objects in the current session. As this lesson was prepared, the data are saved in a separate file, the syntax has been prepared using an ASCII text editor and then saved in a separate file, etc. As such, everything in this session can be recreated easily if needed. Of course, the use of rm() should only be used after careful thought and assurance that adequate measures have been taken to save and backup all data and syntax.

To reduce even more clutter and to avoid possible conflict in redundant function names in different packages, it may be helpful to detach some (or all) attached packages. After all, they can always be put back into use later, when needed again. The detach() function is possibly the best way to detach these packages. First, confirm the library packages currently loaded into the R session and their location.

```
.libPaths()    # Identify the library pathname.
.Library       # Identify the library pathname.

search()       # Give information about the current R session.
searchpaths()  # Give information about the current R session.

sessionInfo()         # List all installed packages.
installed.packages()  # List all installed packages.
```

Then, note below how a few of the many available packages are detached.

```
detach("package:epicalc")      # Detach a package.
detach("package:vioplot")      # Detach a package.

sessionInfo()         # List all installed packages.
installed.packages    # List all installed packages.
```

It may instead be best to actually remove some packages if they are simply not needed. The `remove.packages()` function is used to serve this purpose.

```
installed.packages() # List all installed packages.

remove.packages(c("gmodels"))
remove.packages(c("s20x"))

installed.packages() # List all installed packages.
```

Again, it is a matter of personal preference to detach or remove packages that are currently not needed. These housekeeping actions can help reduce conflicts that sometimes happen when two separate packages use the same name for functions that serve different purposes.

```
###############################################
# Prepare to Exit, Save, and Later Retrieve #
# This R Session                             #
###############################################

getwd()             # Identify the current working directory.
ls()                # List all objects in the working
                    # directory.
ls.str()            # List all objects, with finite detail.
list.files()        # List files at the PC directory.

save.image("Two-Way_ANOVA_Formula-and-TimeOfDay.rdata")

getwd()             # Identify the current working directory.
ls()                # List all objects in the working
                    # directory.
ls.str()            # List all objects, with finite detail.
list.files()        # List files at the PC directory.


alarm()             # Alarm, notice of upcoming action.
q()                 # Quit this session.
                    # Prepare for Save workspace image? query.
####################### END #######################
```

Use the R graphical user interface (GUI) to load the saved .rdata file: file and then load Workspace. Otherwise, use the `load()` function, keying the full pathname, to load the .rdata file and retrieve the session. Recall, however, that it may be just as useful to simply use the .R script file and recreate the analyses and graphics, provided the data files remain available.